A LINEAR PROGRAMMING MODEL FOR AIR POLLUTION CONTROL

The MIT Press
Cambridge, Massachusetts, and
London, England

A LINEAR PROGRAMMING MODEL FOR AIR POLLUTION CONTROL

Robert E. Kohn

Partial support for writing this book
was provided under National Science
Foundation Grant SOC–74–07802. The
conclusions contained herein are those
of the author and do not necessarily
reflect the views of the National Science
Foundation.

This book was printed and bound
in the United States of America.

Library of Congress Cataloging in
Publication Data

Kohn, Robert E
   A linear programming model for
air pollution control.

    Bibliography: p.
    Includes index.
    1. Air—Pollution—Linear program-
ming. I. Title. TD883.1.K64
363.6   78–17292
ISBN 0–262–11062–8

To my wife, Martha

CONTENTS

LIST OF FIGURES

LIST OF TABLES

PREFACE

The first major study of air pollution emissions in the United States was undertaken in the St. Louis region as a cooperative effort by federal, state, and local agencies. While still in mimeograph form, this study was made available to me and provided a data base for a linear programming model of air pollution control. The basic structure of that model was developed in my doctoral dissertation.

This research continued for seven years. The original data base was augmented as new aspects of air pollution abatement were examined. Some of the results of this research have been published in articles appearing in journals and conference volumes from 1970 through 1977. Each of these publications is based on a different extension of the model, incorporating additional data. Nowhere have the results of these individual extensions been interrelated; this is one of the objectives of the present volume. Another objective is to incorporate the linear programming model of air pollution control into the context of a general equilibrium analysis, in which the results have welfare economic implications. In this broader perspective the research takes on new meaning and some of the results that had gone unnoticed are found to be important.

The initial research on the linear programming model was supported by a doctoral fellowship from the U.S. Public Health Service. Subsequent funding was provided under three consecutive grants from the National Science Foundation and augmented by various kinds of support from the Business Division, Computing Facilities, and the Office of Research and Projects, all of Southern Illinois University, Edwardsville.

A number of colleagues contributed to the development of this model. These include the editors of journals, their referees, and the discussants who have seen pieces of the model and whose comments have opened up new paths and directions. I am grateful to John Kenneth Gohagen of Washington University for his in-depth review of this manuscript. His suggestions and his influence are reflected in every chapter. I have also benefited from the help of David Ault, Donald Aucamp, David Harrison, Jr., Jerome Hollenhorst, Karlyn Klopmeier, Richard Parker, and other colleagues. The students in my graduate economics seminar and in the environmental studies program at Southern Illinois University used a preliminary version of this book as a text and provided a number of useful ideas. I am grateful to Gloria Horkits, Virginia Schneider, and Lisa Wyatt for typing successive drafts with enduring patience. Finally, I thank my wife Martha for her support and encouragement.

This book is written for economists, air pollution control planners, and for engineers and mathematicians who are interested in the application of linear programming to social problems. For readers who are planning to implement similar models in other airsheds, this book anticipates many of the problems they are likely to encounter and illustrates the richness of results that can be obtained.

A LINEAR PROGRAMMING MODEL FOR AIR POLLUTION CONTROL

1

THE LINEAR
PROGRAMMING MODEL

The major problems of air pollution occur as a consequence of the clustering of polluting activities. Emissions from many sources intermix and accumulate within geographic spaces that we call airsheds. The problem is one of reducing emissions to achieve desired air quality.

Three themes appear in this book. They are (1) airshed planning, (2) simulation of pollution abatement, and (3) economic efficiency. The method of linear programming provides a powerful tool for air pollution control planning at the local level. The usefulness of this method is demonstrated in this book by a score of empirical applications, based on data for the St. Louis airshed. Many of these results are sufficiently general that they are applicable to other airsheds, as well as to pollution control efforts that are not necessarily least-cost. In effect, then, this research provides a simulation model for examining some important issues in air pollution control.

The Linear Programming Model (which is now capitalized to denote this particular application to air pollution control) is a submodel of a general equilibrium model developed in this chapter. It is useful to view an airshed as an independent economy in which the utility levels of the inhabitants are maximized. The tools of welfare economics can then be applied to determine conditions for an efficient allocation of resources into production and abatement activities. It follows that abatement strategy should be formulated at the airshed level and should comprise a least-cost set of pollution control activities.

The first sections of this chapter are introductory in nature. The three themes of airshed planning, simulation, and economic efficiency are discussed. Following this, the mathematical analysis is presented in which we progress from a general equilibrium model to the Linear Programming Model for Air Pollution Control.

Airshed Planning

It is frequently the case that individual airsheds are politically fragmented, and different regulations and policies are enforced in the separate jurisdictions of the same airshed. Because emissions intermix, it is crucial that planing should take place at the airshed level.

Several planning models are presented in this book. The simpler one is a version of the Linear Programming Model in which it is assumed that the ambient air concentration of a pollutant, over and above the background concentration, is proportional to total emissions of that pollutant. Although this assumption on total emissions has limitations, it is the basis of many federally sponsored strategies, such as Inspection/Maintenance (of automobiles) and the Offset Rule (for new sources in an airshed).

An alternative version of the Linear Programming Model is based on a diffusion formula in which the geographic locations of individual pollution sources are taken into account. In this version of the model, abatement strategy includes the selective location of new sources and even the relocation of existing sources. Some useful information, based on the empirical application of this model, is presented in chapters 4 and 7.

Both types of planning models have their place. The model based on total emission flows, however, is less complex and can be more easily implemented by regulatory officials. Because local agencies are required to maintain emission inventories, the data base is easily kept current. For these reasons, the simpler model is a recommended planning tool.

In general, there is no simple trade-off between money spent on abatement and the level of air quality. Air quality is based on a number of different pollutants, and there are trade-offs between them. The Linear Programming Model is essential when there are multiple requirements. This is illustrated in chapter 3 by the example of low sulfur coal. For many years, pollution control in St. Louis was hampered by legal controversy over regulations that specified the maximum sulfur content of coal. Other pollutants in addition to sulfur dioxide were affected, and the controversy was complicated by side issues such as the scarcity of natural gas, which might be substituted for coal. With the Linear Programming Model the various trade-offs were put into proper perspective.

The solution of the Linear Programming Model shows how an entire set of air quality standards can be achieved at the least total cost of abatement. Furthermore, the sensitivity of the solution to each of the control method costs can be readily determined. This allows greater flexibility in using the model. For example, the enforcement of a particular abatement activity might appear inequitable to control officials; if a more acceptable control activity were inefficient by a small percents of its unit cost, some trade-off of efficiency for equity might be justified. The power of linear programming is greatly enhanced by various capabilities for sensitivity analysis.

The emphasis here is on a model to assist in planning air quality strategy. New problems inevitably confront policy makers. In one year it makes sense to promote the conversion of coal furnaces to natural gas. In a later year the prevailing sentiment may be for conserving natural gas and converting back to coal. Whether the control agency should alter its earlier policy could depend on the magnitude of the increased cost of pollution abatement per cubic foot of gas to be replaced by coal. This cost estimate is easily obtained with the Linear Programming Model. Similarly, the emphasis that should be placed on reducing automotive emissions by Inspection/Maintenance or by car pooling can be given a dollar value in saved abatement costs. Given the many conflicting objectives in air pollution control, regulatory officials could use this planning tool to expand their awareness and to reinforce or modify their own intuitive judgments.

Simulation of Pollution Abatement

The Linear Programming Model is implemented with data for the St. Louis airshed. The data presented and discussed in chapter 2 underlie a series of models that are described in this book. These are listed in table 1.1. Although the table includes only thirteen models, some of them have several sub-versions, so that there are in fact more than twenty models.

In total this represents an extensive simulation of the economics of air pollution control. In many cases, issues are resolved that would have continued to trouble policy makers. Some of the questions examined in this book and identified by the corresponding model number in the table are as follows:

Efficiency savings. Aside from regulatory and enforcement costs, how great is the saving in total abatement cost for the efficient solution as compared to the current regulatory solution in the St. Louis airshed? (I)

Joint-wastes. Is it necessary to include all forms of wastes in an environmental model? Is there a danger that a planning model for air pollution alone will yield a solution that would seriously augment the flows of solid, liquid, and thermal wastes (which occur as a consequence of the cessation of open burning, increased use of scrubbers, generation of electricity to operate control equipment, etc.)? (II)

Cost of confidence. The relationship between emissions and ambient air concentrations is affected by stochastic meteorological variables such as wind velocity. What is the risk in using average values for these stochastic

Table 1.1
Major versions of the Linear Programming Model

Model Number	Chapter	Description
I	2, 3	The basic model with fixed pollution source levels and maximum allowable total annual emission flows.
II	3	This model is identical to Model I except that it includes output coefficients for liquid, thermal, and solid wastes as well as air pollutants.
III	4	This model is the same as Model I except that it incorporates the Larsen formula relating total emissions and annual average concentrations. Accordingly, maximum pollutant concentrations can be entered directly into the model.
IV	4	This is an elaboration of Model III, in which the linear relationship between total emissions and pollutant concentrations is stochastic. A specific probability that the desired air quality goals will be achieved is a parameter of this model.
V	4	A diffusion formula relates emissions of each source to pollutant concentrations at the CAMP Station. Composite sources are disaggregated according to their location in the airshed. With Model V, the location of any point source can be varied and the consequent effect on pollutant concentrations and abatement costs thereby determined.
VI	4	An alternative version of Model I in which source magnitudes are projected for 1985. The total allowable pollutant flows for 1985 are the same as those in Model I.
VII	5	This is a benefit-cost version of Model III. In place of maximum pollutant concentrations, there is an objective function equal to the dot product of pollutant concentrations (now variables) and their respective shadow prices from Model III, plus the total cost of abatement. The inclusion of unit capital coefficients for the abatement activity variables allows for alternative rates of interest.
VIII	5	A benefit-effectiveness version of Model III in which a pollution index is minimized subject to a constraint on the total cost of abatement. In the two-pollutant case, the concentration of particulates is minimized for a range of sulfur dioxide concentrations, with the concentrations of the other pollutants and the total cost of abatement held constant.
IX	6	A version of Model I incorporating the feedback of abatement activities on pollution source magnitudes. As a consequence of abatement, the selling prices of pollution related goods are higher, and the quantities demanded are accordingly reduced.

Model Number	Chapter	Description
X	6	An extension of Model IX, allowing for voluntary substitutions of natural gas to avoid costly restrictions on coal burning.
XI	6	In this version of Model I, the feedback of abatement activity on pollution source magnitudes occurs through the production of inputs for abatement.
XII	6	An extension of Model XI that includes a matrix of input-output multipliers and accounts for both the direct and the indirect demand for inputs for abatement.
XIII	7	An extension of Model IX in which job displacement, caused by reductions in output, is measured. In addition, there is a direct loss of jobs as a consequence of abatement technology. In an equity version of this model, job displacement is minimized subject to a constraint on the total cost of abatement.

variables, and how great is the cost of insuring against this risk by more intensive permanent abatement? Is this cost so high as to justify greater reliance on episode control measures? (IV)

Locational selectivity. How important is the selective location of new sources as a strategy for pollution control? (V)

Abatement and industrial growth. If total emissions are held to fixed allowable levels, is this likely to seriously restrain economic growth in an airshed? (VI)

Benefit-effectiveness. Are air quality standards in the St. Louis airshed benefit-effective, or should the standards for certain pollutants be made more stringent? (VIII)

Substitution effects. The costs of abatement are likely to increase prices of pollution-related goods and induce some substitutions of other goods. Will these substitutions significantly improve air quality and should they be reflected in regulatory planning? (IX)

Derived demands. The abatement of pollution requires inputs whose production is itself polluting. To what extent, therefore, must an abatement effort be augmented to offset this feedback effect? (XI)

Employment impact. Should regulatory agencies adjust their control strategies to preserve existing jobs? (XIII)

These are some of the issues that are examined empirically in this book. In addition, the Linear Programming Model is used to simulate the revenue potential of a program of pollution fees and the capability of such a program for internalizing the pollution costs of land use. The results of the various simulation models are interpreted in this book. This aspect of the research should be useful to economic theorists as well as regulatory policy makers. Furthermore, it demonstrates the wide versatility of the Linear Programming Model.

Economic Efficiency

In the economic literature on pollution control in an airshed, there are two major theoretical approaches. One is typified by the *general equilibrium model*, in which prices and outputs of individual goods are variable. Such models are useful in defining optimal levels of environmental quality and quantities of private goods. The second approach is *partial equilibrium analysis*, in which prices are essentially fixed and pollution control is accomplished without any changes in the quantities of goods and services associated with polluting activities. While this approach facilitates empirical research, the underlying assumption is faulty; pollution abatement, through price, income, and derived demand effects, does alter the levels of pollution-related outputs.

There has been some effort by economists to reconcile the two approaches to environmental analysis. Dick (1974, p. 125) constructs a partial equilibrium model in which the pollution from a single industry is being controlled, while the outputs and pollution flows from all other industries are assumed to be optimal to begin with. A similar assumption is implicit in the partial equilibrium models of Dorfman (1972, pp. xvi–xviii) and of Meade (1973, p. 58). It is more characteristic of the real world, however, that an entire spectrum of polluting activities is not properly controlled to begin with, and a strategy for simultaneous abatement by all sources is desired.

In the mathematical section of this chapter, we develop a general equilibrium model that has the useful properties of a partial equilibrium model. This is achieved with a simplifying assumption that air pollution originates exclusively in the production of intermediate goods, for which quantities demanded are proportional to some resource input. Assuming that resources are fixed, this model is one in which prices and outputs of the final goods are variable, but the magnitude of each pollution source is fixed by virtue of the assumption of technology. Furthermore, we assume away the relative price effects of abatement, as well as the real income effects, which would otherwise alter pollution source levels.

The Linear Programming Model of air pollution control is a component of the aforementioned model. Although the simplifying assumptions on pollution and intermediate activities are artificial, they permit us to relate the results of the Linear Programming Model to some important conditions for economic efficiency. For example, the shadow prices of the individual pollutant standards must be equal to (or less than) the marginal benefits of abatement for the respective pollutants. It follows that there is an optimal concentration for each pollutant. This analysis is pursued in chapter 5, where the economic efficiency of the pollutant concentrations is empirically tested.

In welfare economic theory, efficiency and equity are separate policy considerations. The level of air quality that is optimal under one distribution of income may not be optimal under a different distribution of income. It is appropriate, therefore, that regulatory agencies be cognizant of equity as well as cost-effectiveness. In chapter 7 alternative governmental programs for pollution control are evaluated with respect to efficiency and equity criteria, and it is found that there is an overlapping of the two objectives. For this evaluation of governmental programs, the empirical data from the Linear Programming Model are useful.

Although the basic planning model is based on the assumption that pollution source levels are constant, this assumption is dropped in chapter 6. In that chapter the Linear Programming Model is adapted to the case in which abatement increases the market prices of goods and, as a consequence of estimated reductions in quantities demanded, decreases pollution source levels. The effect of this feedback is to reduce the total cost of achieving a given set of air quality standards. Alternatively, pollution control requires inputs for abatement. The derived demand for these inputs increases pollution source levels, and the effect of this feedback is to increase the total cost of achieving a given set of standards. The models in chapter 6 more closely approximate the economic interactions that are characteristic of general equilibrium analysis.

Mathematical Analysis

In the remainder of this chapter mathematical models are developed, that lead ultimately to the Linear Programming Model for Air Pollution Control. The reader who plans to follow the mathematical derivations is urged, at the outset, to become acquainted with the Glossary of Mathematical Symbols that follows the Appendix.

The sequence of models that complete this chapter are as follows. We first examine a general equilibrium model in which there are two goods and one pollutant. Here pollution originates during the production of one of the goods and is undesirable because it diminishes the utility of both households. The thrust of the model is that an excessive level of pollution will be reduced by (1) adoption of alternative processes of production that are less polluting per unit of output, (2) shifts in consumption away from the good that is polluting in production, and (3) a contraction in units of total output.

Next we develop a pure abatement model in which economic efficiency is achieved solely by abatement, not by shifts in the composition of final outputs nor by a contraction of units of output. Such a model is more closely related to conventional linear programming models of air pollution control, in which the outputs of polluting activities are assumed fixed. This model is extended to include more than one air pollutant and is reformulated as a linear programming problem. The chapter concludes with an appendix that illustrates the equivalence of emission standards and pollution fees in the case of the pure abatement model.

The General Equilibrium Model

Consider a simple economy consisting of two households, household-one and household-two. We shall assume that the labor supplied by these households is the sole factor of production and that the quantity supplied is perfectly inelastic. In this economy two goods are manufactured, good-one and good-two, and their quantities are expressed by the variables y_1 and y_2. The allocations to the households are the sets (y_{11}, y_{21}) and (y_{12}, y_{22}), where the second subscript denotes the consuming household. The quantities produced are entirely consumed:

$$y_{11} + y_{12} = y_1,$$
$$y_{21} + y_{22} = y_2. \tag{1.1}$$

We shall assume that the production of good-one, but not of good-two, is polluting. The activity level of the least-cost process for making good-one is the variable x_{1a}, and the corresponding rate of emissions is e_{1a}. The technology of abatement is formulated in terms of alternative processes, which are less polluting but more costly. Assuming that there are four processes 1a, 1b,

1c, and 1d for making good-one and measuring their activity levels x_{1a}, x_{1b}, x_{1c}, and x_{1d}, in units of good-one produced it follows that

$$x_{1a} + x_{1b} + x_{1c} + x_{1d} = y_1. \tag{1.2}$$

The alphanumeric ordering of subscripts is such that

$$
\begin{aligned}
e_{1a} &> e_{1b} > e_{1c} > e_{1d}, \\
c_{1a} &< c_{1b} < c_{1c} < c_{1d},
\end{aligned} \tag{1.3}
$$

where e_j is the emission rate, expressed in parts per million (or micrograms per cubic meter) per unit of activity-j, and c_j is the cost coefficient measured in labor units. If the activity of the nonpolluting process for making good-two is y_2 and the unit cost is c_2, it follows that

$$
\begin{aligned}
c_{1a}x_{1a} + c_{1b}x_{1b} + c_{1c}x_{1c} + c_{1d}x_{1d} + c_2 y_2 &= R, \\
e_{1a}x_{1a} + e_{1b}x_{1b} + e_{1c}x_{1c} + e_{1d}x_{1d} &= q,
\end{aligned} \tag{1.4}
$$

where R is the fixed supply of labor and q is the level of air pollution.

According to welfare economic theory, a vector of outputs, (y_{11}^*, y_{12}^*, y_{21}^*, y_{22}^*, q^*), is optimal if there is no alternative attainable set of outputs that would make one of the households better off without making the other household worse off. Better or worse off for a household is measured according to a utility function for that household. Thus

$$U^i(\bar{y}_{11}, \bar{y}_{21}, \bar{q}) > U^i(\tilde{y}_{11}, \tilde{y}_{21}, \bar{q}) \tag{1.5}$$

is equivalent to the statement that the ith household is better off[1] with the combination ($\bar{y}_{11}, \bar{y}_{21}, \bar{q}$) than with ($\tilde{y}_{11}, \tilde{y}_{21}, \bar{q}$). We shall make the conventional assumptions that utility increases at a decreasing rate as the quantity of a private good increases, and decreases at an increasing rate as the level of air pollution increases.

The welfare economic conditions for an efficient allocation of outputs, for example one in which $U^2 \geqslant \bar{U}^2$ and U^1 is maximized, are derived from a Lagrangian expression incorporating (1.1), (1.2), and (1.4):

$$
\begin{aligned}
\mathscr{L} = \; & U^1(y_{11}, y_{21}, q) + \lambda_u[U^2(y_{12}, y_{22}, q) - \bar{U}^2] \\
& + \lambda_r[R - c_{1a}x_{1a} - c_{1b}x_{1b} - c_{1c}x_{1c} - c_{1d}x_{1d} - c_2(y_{21} + y_{22})] \\
& + \lambda_1(y_{11} + y_{12} - x_{1a} - x_{1b} - x_{1c} - x_{1d}),
\end{aligned} \tag{1.6}
$$

where $q = e_{1a}x_{1a} + e_{1b}x_{1b} + e_{1c}x_{1c} + e_{1d}x_{1d}$. Observe that q, without a subscript, enters both utility functions, implying that the two households are exposed to the same level of pollution. (The assumption of equal exposure can be dropped without altering the essential results. See Kohn, 1975b, pp. 26–28). Because exposure by one household does not change the level to which the other household is exposed, the level of air pollution is analogous to a pure public good as conceived by Samuelson (1954).

We shall examine the case in which the solution of (1.6) is one in which each household consumes both goods and a combination of processes 1b and 1c is optimal.

From the Kuhn-Tucker conditions for optimality,

$$y_{ik}(\partial L/\partial y_{ik}) = 0,$$
$$x_j(\partial L/\partial x_j) = 0,$$
(1.7)

it follows that

$$- (U_q^1/U_2^1 + U_q^1/U_2^2) = \frac{\left(\dfrac{c_{1c} - c_{1b}}{e_{1b} - e_{1c}}\right)}{c_2},$$
(1.8)

$$U_1^1/U_2^1 = U_1^2/U_2^2 = \frac{c_{1b} + e_{1b}\left(\dfrac{c_{1c} - c_{1b}}{e_{1b} - e_{1c}}\right)}{c_2},$$
(1.9)

where U_j^i/U_k^i is the ith household's marginal rate of substitution of good k for good j (or air quality q). Equation (1.8) states that the quantities of good-two which the two households together would exchange for a unit decrease in air pollution must equal the opportunity cost to the producing sector, as measured in units of good-two, of abating one unit of pollution, holding the output of good-one constant. The right-hand-side value of (1.8) is the ratio of the marginal cost of abatement, $(c_{1c} - c_{1b})/(e_{1b} - e_{1c})$, to the marginal cost of good-two.[2] For convenience we shall henceforth assume that the wage rate is one dollar so that all costs can be expressed in dollar values. If (1.8) is multiplied through by c_2, the left-hand side becomes the marginal benefit of abatement, measured in dollars, and the right-hand side the marginal cost of abatement. The equality of marginal benefits and costs is a well-known condition for an optimal supply of a public good.

Equation (1.9) states that the marginal rate of substitution in consumption between good-one and good-two is the same for both households and is equal to the rate of transformation in production, holding the pollution level constant. The rate of transformation between the two goods is the ratio of their marginal costs. Whereas the marginal cost of good-two is simply the direct cost c_2, the marginal cost of good-one, holding the level of air pollution constant, is $c_{1b} + e_{1b} [(c_{1c} - c_{1b})/(e_{1b} - e_{1c})]$. This is the direct cost of labor, c_{1b}, plus the cost of eliminating the incremental pollution, e_{1b}. An equivalent expression for the marginal cost of good-one is $c_{1c} + e_{1c} [(c_{1c} - c_{1b})/(e_{1b} - e_{1c})]$.

An allocation that fulfills the above conditions is illustrated graphically by the point P in figure 1.1. This point is on th facet bce of the simplex abcde. Observe that vertices a, b, c, and d are on the q-y_1 plane and vertex e is on the q-y_2 plane. For expositional convenience the q-axis begins at q_0, which is the maximum level of pollution for this economy. Thus more desirable levels of all three outputs are outward from the origin. Points on the facet abe represent sets of q, y_1, and y_2 obtainable with combinations of processes 1a, 1b, and 2. Points on the facet bec are obtained with combinations of processes 1b, 1c, and 2. The final facet cde is generated by combinations of processes 1c, 1d, and 2. The frontier in figure 1.1 is convex because all four processes for making good-one are technically efficient. This is the case because

$$\frac{c_{1b} - c_{1a}}{e_{1a} - e_{1b}} < \frac{c_{1c} - c_{1b}}{e_{1b} - e_{1c}} < \frac{c_{1d} - c_{1c}}{e_{1c} - e_{1d}}. \tag{1.10}$$

Whereas P denotes the vector (y_1^*, y_2^*, q^*), points A^1 and A^2, which are inside the simplex (see upper right-hand side inset of figure 1.1), denote the sets $(y_{11}^*, y_{21}^*, q^*)$ and $(y_{12}^*, y_{22}^*, q^*)$, respectively. Geometrically, A^1 and A^2 are vertices of a parallelogram $A^1 P A^2 q^*$ that lies on a plane parallel to the y_1-y_2 axis through q^*. This plane contains the coordinate axes in the right-hand inset of figure 1.1. The attained indifference curve of each household is superimposed on the same coordinate system, with consumption by each household measured independently along the corresponding axes. The slopes of the indifference curves $\partial y_{21}/\partial y_{11}$ at A^1 and $\partial y_{22}/\partial y_{12}$ at A^2 are equal to each other and to the slope $\partial y_2/\partial y_1$ of the facet through P, thereby satisfying condition (1.9). Condition (1.8) will be graphically illustrated in a comparable model later in this chapter.

If we assume that the two goods in this economy are sold in perfectly competitive markets, that households do not consider the pollution-generating consequences of their consumption decisions, and there is no government program for pollution control, producers of good-one would use the least-cost

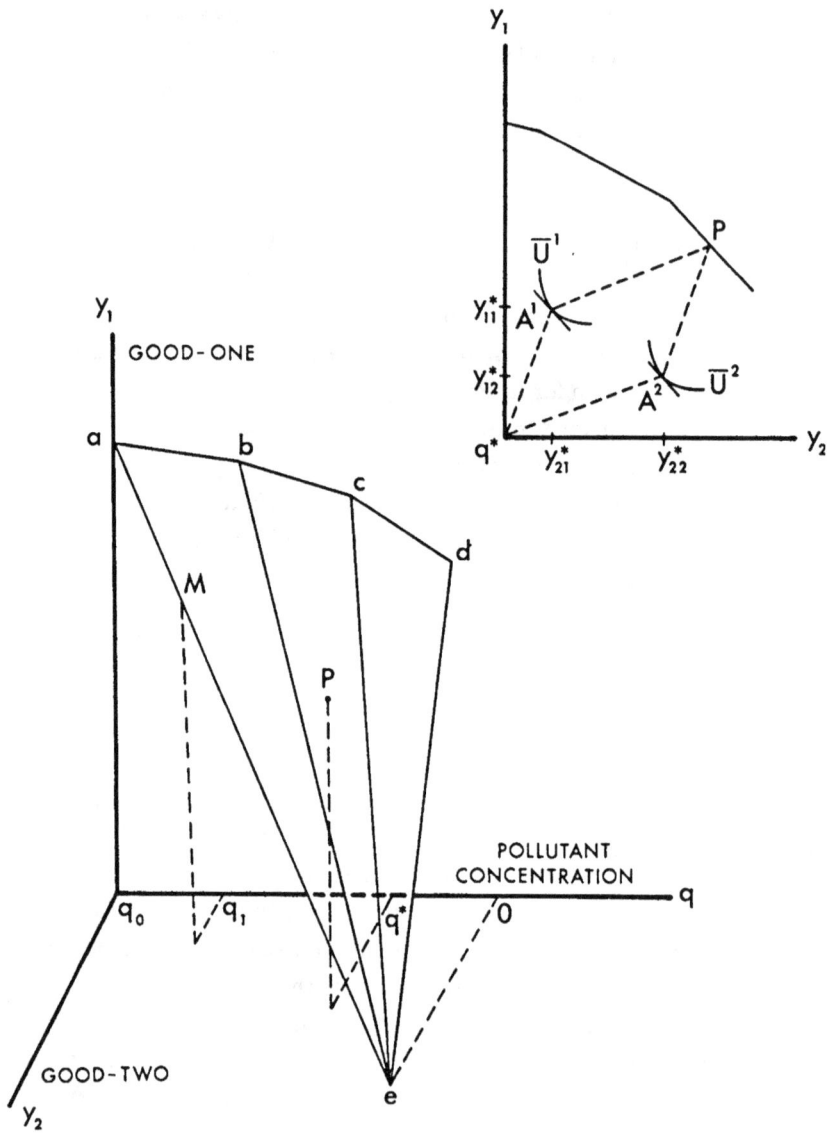

Figure 1.1
The production-possibility frontier and an optimal allocation

process, and the selling prices of the two goods would be c_{1a} and c_2, respectively. The competitive market allocation would correspond to a point such as M in figure 1.1, and the level of pollution would be q_1.

To reduce the pollution level from q_1 to q^*, three things generally happen: (1) technological abatement, (2) a shift in the ratio of goods consumed, and (3) a contraction in total units of output. Condition (1.8) implies that a combination of processes 1b and 1c should be used for making good-one. Because emission rates for this combination of processes are less than e_{1a} there would be a decrease in the level of pollution. Condition (1.9) implies that consumption decisions should be based on a price for good-one that is higher than the initial price, c_{1a}. In general, a higher relative price for good-one will result in some substitution of good-two, which is nonpolluting in production, for good-one. In addition, the increased cost of production (c_{1b} exceeds c_{1a}) reduces the quantity of goods that can be produced with the fixed supply of labor. This further decreases the output of good-one and hence the total flow of emissions. The three effects combine to reduce pollution from q_1 to q^*.

In theory this result would be achived in a perfectly competitive economy by assessing a pollution fee equal to $(c_{1c} - c_{1b})/(e_{1b} - e_{1c})$ per unit of pollution emitted. Producers of good-one would minimize total costs of production (which includes pollution fee charges) by using a combination of process 1b and 1c. There is the problem of divisibility here (see Kohn, 1975b, pp. 81–86) in that *any* combination of these two processes will result in the same total cost of producing a given quantity of good-one. We shall simply assume that the combination of processes 1b and 1c is chosen such that the resulting equilibrium market allocation is one in which condition (1.8) is satisfied. The marginal cost of good-one will be $c_{1b} + e_{1b} [(c_{1c} - c_{1b})/(e_{1b} - e_{1c})]$, which is the sum of direct costs c_{1b} plus pollution fees per unit of output. The marginal cost of good-two will be c^2. Assuming perfect competition, goods will be priced at marginal cost and condition (1.9) satisfied as a consequence of utility maximization by consumers. For the economy to be at full employment, the government must transfer the fee revenue paid by producers, totalling $q^*[(c_{1c} - c_{1b})/(e_{1b} - e_{1c})]$, to the households.[3] These lump sum transfers can, in theory, be given in such ratios as to satisfy the constraint in (1.6) on the utility level of household-two.

If pollution control were accomplished by emission standards, conditions

(1.8) and (1.9) would not be satisfied. Even though the emission standards require producers of good-one to use the optimal combination of process 1b and 1c, the allowable emission flows would cause reductions in utility levels that would not be reflected in the relative price of good-one. In the next section we develop a model in which economic efficiency can be achieved by emission standards as well as by emission fees.

The Pure Abatement Model

The construction of economic models of air pollution control is inevitably complex. There are numerous sources of pollution in an airshed, and it is a formidable task to identify all of these sources and the corresponding alternative production-abatement processes. To further account for the effect of abatement costs on individual output levels would be enormously complicating. It is therefore common for model builders to assume that pollution-related output levels are fixed and independent of abatement. Thus, for example, it is generally assumed in such models that households will demand a specific quantity of electricity, regardless of the cost of abatement at the power plant; or if the substitution of natural gas for coal is an abatement alternative for specific industrial furnaces, that the same total heat will be required with either fuel.

Accordingly, we revise our general equilibrium model so that polluting activity levels are in fact independent of abatement costs. This result is obtained by assuming that the good that is polluting in production is an intermediate good, used by firms in a fixed proportion to their labor input. Four alternative processes may contribute to the quantity of this intermediate good s_1:

$$x_{1a} + x_{1b} + x_{1c} + x_{1d} = s_1. \tag{1.11}$$

There are two final goods, y_2 and y_3, produced by nonpolluting processes, y_2 and y_3. The total resource requirements in this economy are

$$c_{1a}x_{1a} + c_{1b}x_{1b} + c_{1c}x_{1c} + c_{1d}x_{1d} + c_2 y_2 + c_3 y_3 = R. \tag{1.12}$$

All firms, including those which manufacture the intermediate good, require a quantity of the intermediate good proportional to their labor requirements. This proportion α is the same for all firms.[4] Thus

$$\alpha c_{1a}x_{1a} + \alpha c_{1b}x_{1b} + \alpha c_{1c}x_{1c} + \alpha c_{1d}x_{1d} + \alpha c_2 y_2 + \alpha c_3 y_3 = s_1. \tag{1.13}$$

It follows from (1.12) and (1.13) that

$$\alpha R = s_1 \tag{1.14}$$

and that the quantity of the polluting output s_1 is indeed fixed. That abatement activity itself should require inputs that give rise to pollution was originally suggested by Leontief (1970).

The welfare economic conditions for an efficient allocation of inputs and outputs are derived from the following Lagrangian expression, which incorporates (1.11), (1.12), and (1.13):

$$
\begin{aligned}
\mathcal{L} = {}& U^1(y_{21}, y_{31}, q) + \lambda_u[U^2(y_{22}, y_{32}, q) - \bar{U}^2] \\
& + \lambda_r[R - c_{1a}x_{1a} - c_{1b}x_{1b} - c_{1c}x_{1c} - c_{1d}x_{1d} - c_2(y_{21} + y_{22}) \\
& - c_3(y_{31} + y_{32})] + \lambda_1[\alpha c_2(y_{21} + y_{22}) + \alpha c_3(y_{31} + y_{32}) \\
& - (1 - \alpha c_{1a})x_{1a} - (1 - \alpha c_{1b})x_{1b} - (1 - \alpha c_{1c})x_{1c} - (1 - \alpha c_{1d})x_{1d}],
\end{aligned}
\tag{1.15}
$$

where $q = e_{1a}x_{1a} + e_{1b}x_{1b} + e_{1c}x_{1c} + e_{1d}x_{1d}$. For the case in which the optimal solution to (1.15) is one in which some of each good is consumed by both households and the intermediate good is produced by a combination of processes 1b and 1c, the conditions for economic efficiency include the following:

$$
-(U_q^1/U_2^1 + U_q^2/U_2^2) = \frac{\left(\dfrac{c_{1c} - c_{1b}}{e_{1b} - e_{1c}}\right)}{c_2}, \tag{1.16}
$$

$$
U_2^1/U_3^1 = U_2^2/U_3^2 = c_2/c_3. \tag{1.17}
$$

Condition (1.16) is identical to (1.8). Condition (1.17), which differs from (1.9), indicates that the optimal marginal rate of substitution is independent of the level of pollution or of abatement. Although both final goods are indirectly polluting, the pollution is proportional to their production costs, and their relative prices do not change. There is a real income effect of technological abatement, causing consumption of final goods to decrease.

In this model, pollution control is accomplished entirely by technological abatement. The output of the intermediate good, which is the source of air pollution, is constant. Shifts in consumption between the final goods and contractions in their total output have no additional impact on the level of pollution. The properties of the Pure Abatement model are such that economic efficiency can be achieved either by emission standards or by emission fees.

The Linear Programming Model 15

Furthermore, economic efficiency can be interpreted in terms of a composite good.

Composite Good

The production-possibility frontier is illustrated in figure 1.2. Each facet, generated by a combination of four processes, is an equilateral trapezoid. The slope $\partial y_2/\partial y_3$ is equal to c_3/c_2 on each of the facets. This is a consequence of the Pure Abatement model in which relative prices of final goods are constant. Hicks (1965, p. 33) has noted that "A collection of physical things can always be treated as if they were divisible into units of a single commodity so long as their relative prices can be assumed to be unchanged in the particular problem at hand." Consequently, this single commodity can be treated as an argument in utility functions. Thus we may define a composite good that is equivalent in cost to, say, $1/c_2$ units of good-two and $1/c_3$ units of good-three, and whose total quantity is $\Upsilon = \Upsilon_1 + \Upsilon_2$.

With this simplification, the Lagrangian expression, becomes

$$
\begin{aligned}
\mathscr{L} = {} & U^1(\Upsilon_1, q) + \lambda_u[U^2(\Upsilon_2, q) - \bar{U}^2] \\
& + \lambda_r[R - c_{1a}x_{1a} - c_{1b}x_{1b} - c_{1c}x_{1c} - c_{1d}x_{1d} - \Upsilon_1 - \Upsilon_2] \\
& + \lambda_1[\alpha(\Upsilon_1 + \Upsilon_2) - (1 - \alpha c_{1a})x_{1a} - (1 - \alpha c_{1b})x_{1b} \\
& - (1 - \alpha c_{1c})x_{1c} - (1 - \alpha c_{1d})x_{1d}],
\end{aligned}
\tag{1.18}
$$

where $q = e_{1a}x_{1a} + e_{1b}x_{1b} + e_{1c}x_{1c} + e_{1d}x_{1d}$. The condition of interest for the case in which x_{1b} and x_{1c} are nonzero and both households consume private goods is

$$
- (U_q^1/U_\Upsilon^1 + U_q^2/U_\Upsilon^2) = (c_{1c} - c_{1b})/(e_{1b} - e_{1c}).
\tag{1.19}
$$

This condition is illustrated in figure 1.3. The set of production possibilities between air pollution and the composite good is represented by the frontier abcd. The level of utility $\bar{U}^2$ allowed household-two is attainable by combinations of outputs along the indifference curve $\bar{U}^2$. The consumption possibilities left for household-one are the vertical distances between the frontier abcd and the indifference curve $\bar{U}^2$ (because both households have the same exposure). These distances are denoted by the MM' curve in figure 1.3.

The utility of household-one is maximized by the combination of outputs at which the MM' curve is tangent to an indifference curve of that household. In figure 1.3 there is a tangency at (q^*, Υ_1^*). Because the slope of the

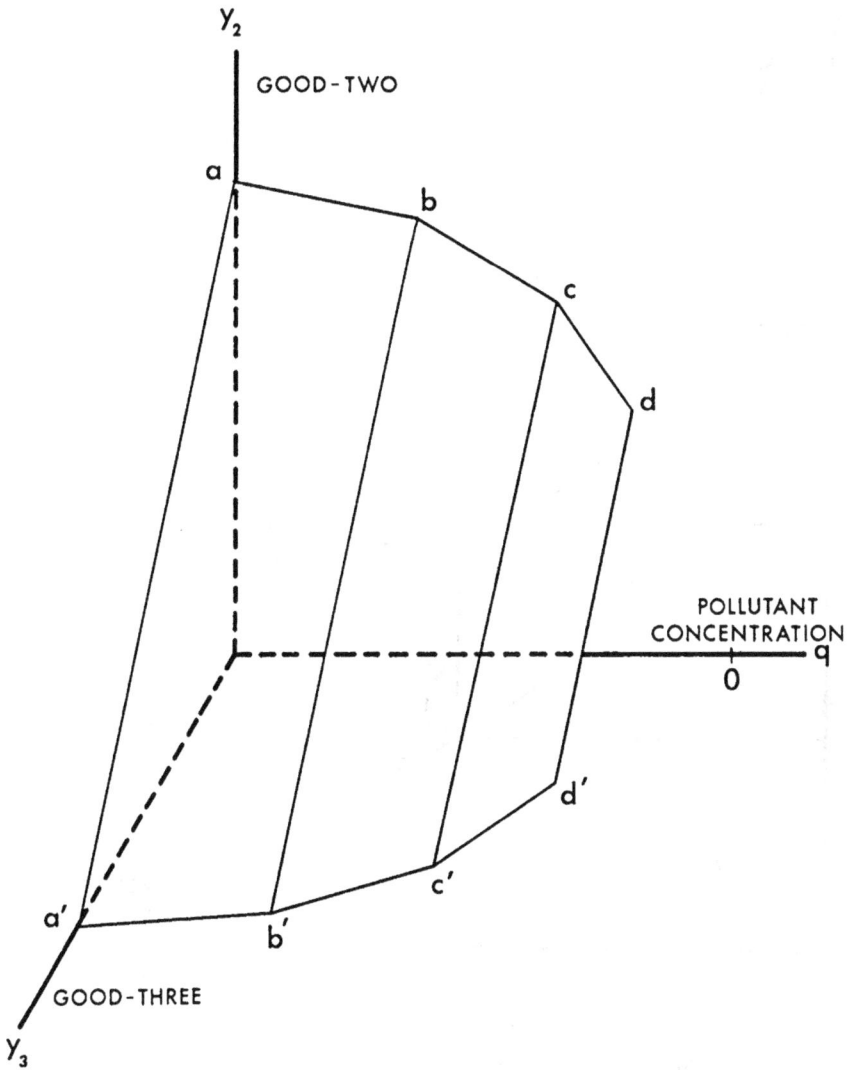

Figure 1.2
Production-possibility frontier for the case in which the polluting good in production is an intermediate good used by all firms in some common fixed proportion to their labor input

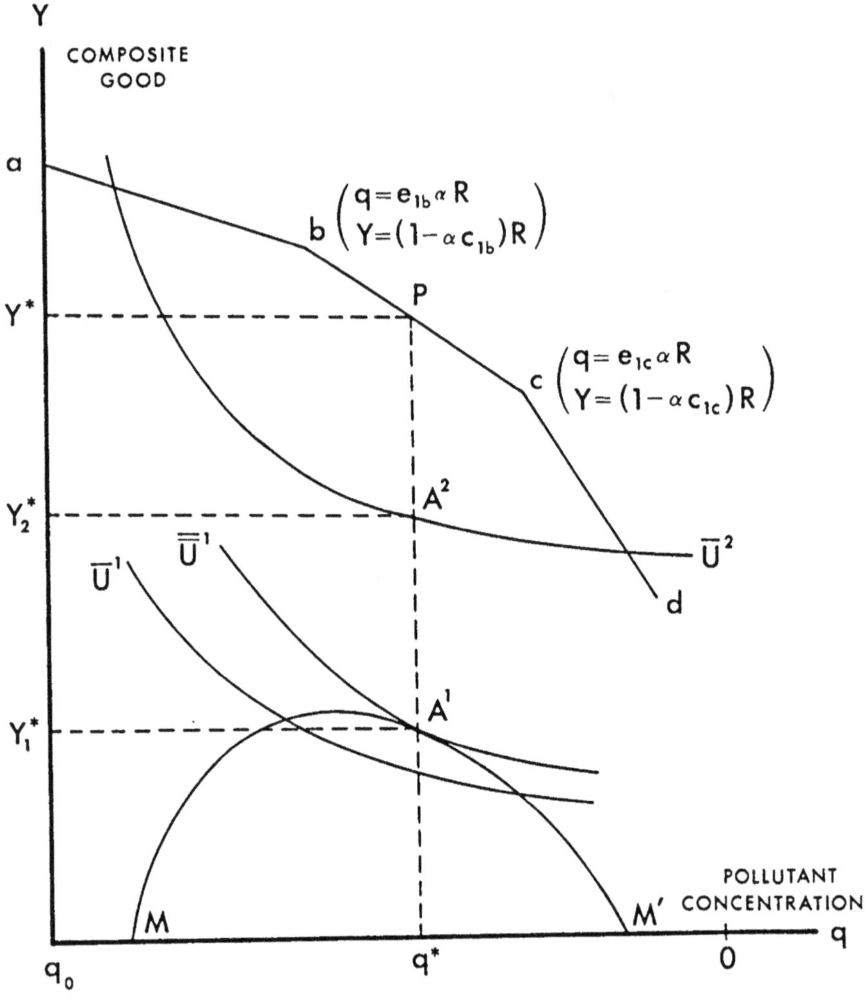

Figure 1.3
An optimal combination of a composite good and air pollution

MM' curve is the difference between the slopes of the frontier abcd and the curve $\bar{U}^2$, it follows that the slopes of $\bar{U}^1$ at A^1 and $\bar{U}^2$ at A^2 sum to the slope of bc; and therefore condition (1.19) is satisfied. The reader may confirm algebraically that the slope of the line segment bc is indeed $(c_{1c} - c_{1b})/(e_{1b} - e_{1c})$.

It is a simple step to generalize the model so that the composite good is a combination of a great many private goods whose relative prices are fixed. The optimal level of air quality will be such that the summation of each household's trade-off of the composite good for air quality will equal the rate of transformation between the two.

The significance of the Pure Abatement model may be summarized as follows: it is based on the assumption that the intermediate good, which is polluting in production, is used by all producers in some fixed proportion to labor cost. The same result would obtain if s_i represented an intermediate production activity engaged in by producers of final goods. It follows that the level of the polluting activity s_i is fixed and the relative prices of final goods do not change because of abatement. The cost of abatement is reflected in reduced purchasing power for final goods. Given a competitive market economy with this technology, economic efficiency can be achieved either by an appropriate set of emission standards or by Pigouvian fees.[5] Furthermore, the Pure Abatement model permits us to represent an entire set of consumer goods by a single composite good.

The Multipollutant Case

The multipollutant case is characteristic of the real world in which emissions of carbon monoxide, hydrocarbons, nitrogen oxides, sulfur dioxide, particulates, benzo(a)pyrene, and other air pollutants may all be associated with a single production process. Here we extend the model of the preceding section to encompass two pollutants whose respective concentrations are q^1 and q^2. The processes for producing the intermediate good generate pollution as follows:

$$e_{1a}^1 x_{1a} + e_{1b}^1 x_{1b} + e_{1c}^1 x_{1c} + e_{1d}^1 x_{1d} = q^1,$$
$$e_{1a}^2 x_{1a} + e_{1b}^2 x_{1b} + e_{1c}^2 x_{1c} + e_{1d}^2 x_{1d} = q^2. \tag{1.20}$$

For a production-abatement process to be considered, it must be the case that at least one of its emission coefficients is less than the emission coefficients (for that same pollutant) for each less costly process. This requirement is less stringent than (1.3) in the one-pollutant case.

The first-order conditions for an optimal combination of the composite good and the two-pollutant concentrations are obtained from the Lagrangian,

$$
\begin{aligned}
\mathscr{L} = {} & U^1(Y_1, q^1, q^2) + \lambda_u[U^2(Y_2, q^1, q^2) - \bar{U}^2] \\
& + \lambda_r[R - c_{1a}x_{1a} - c_{1b}x_{1b} - c_{1c}x_{1c} - c_{1d}x_{1d} - Y_1 - Y_2] \\
& + \lambda_1[\alpha(Y_1 + Y_2) - (1 - \alpha c_{1a})x_{1a} - (1 - \alpha c_{1b})x_{1b} \\
& - (1 - \alpha c_{1c})x_{1c} - (1 - c_{1d})x_{1d}],
\end{aligned} \tag{1.21}
$$

with pollution levels given by (1.20) above. For the case in which both households consume the composite good and x_{1a}, x_{1b}, and x_{1c} are positive, the Kuhn-Tucker conditions yield

$$
(U_{q^1}^1/U_Y^1 + U_{q^1}^2/U_Y^2) = \frac{(c_{1c} - c_{1b})(e_{1a}^2 - e_{1b}^2) - (c_{1b} - c_{1a})(e_{1b}^2 - e_{1c}^2)}{(e_{1a}^1 - e_{1b}^1)(e_{1b}^2 - e_{1c}^2) - (e_{1b}^1 - e_{1c}^1)(e_{1a}^2 - e_{1b}^2)} \tag{1.22}
$$

$$
(U_{q^2}^1/U_Y^1 + U_{q^2}^2/U_Y^2) = \frac{-(c_{1c} - c_{1b})(e_{1a}^1 - e_{1b}^1) + (c_{1b} - c_{1a})(e_{1b}^1 - e_{1c}^1)}{(e_{1a}^1 - e_{1b}^1)(e_{1b}^2 - e_{1c}^2) - (e_{1b}^1 - e_{1c}^1)(e_{1a}^2 - e_{1b}^2)}. \tag{1.23}
$$

These conditions are illustrated in figure 1.4 for the allocations A^1 and A^2. A plane cutting through q^{1*}, parallel to the q^2Y plane, would intersect the convex possibility frontier abcd along two facets. The allocations A^1 and A^2 would lie on a vertical line at q^{2*}, and, as in figure 1.3, the slope of the outward flaring indifference surfaces at A^1 and A^2, that is $U_{q^2}^1/U_Y^1$ and $U_{q^2}^2/U_Y^2$, would sum to the slope $\partial Y/\partial q^2$ of the facet abc. This slope is given by the ratio on the right-hand side of equation (1.23). Likewise, the condition for the optimal allocation of the composite good and pollutant-one could be illustrated on a plane through q^{2*}, parallel to the Yq^1 plane.

The three-dimensional production-possibility frontier in figure 1.4 for two concentrations and a composite good may be compared to the frontier in figure 1.1 for a single concentration and two goods. In the latter the vertices of each facet lie on the $y_1 = 0$ plane and $y_2 = 0$ plane, whereas in the former there are vertices in the interior space. The simplex in figure 1.4 contains three facets, abc, abd, and bcd, whose outer directed normals point in the direction of more Y and lower concentrations of q^1 and q^2. The fourth facet of the tetrahedron, acd, lies under the other three facets and contains technically inefficient combinations of outputs.

There are a number of empirical models in which a single pollutant is

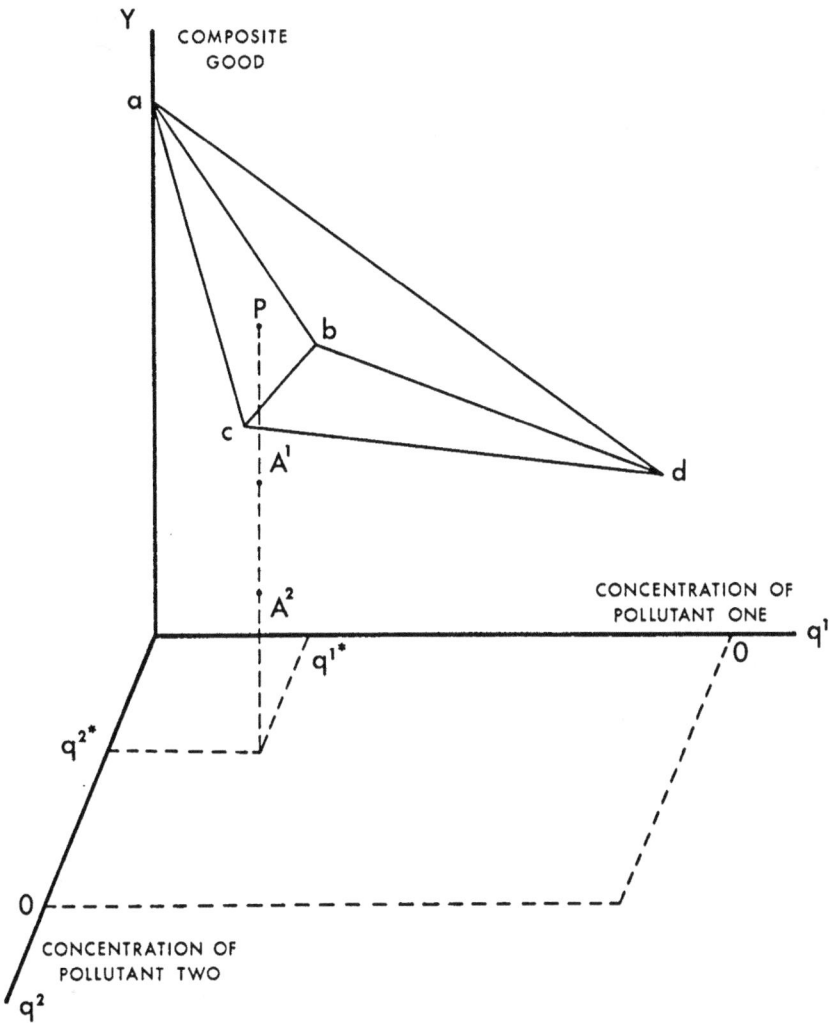

Figure 1.4
Production-possibility frontier for a composite good and two pollutant levels

controlled (see, for example, Kohn, 1968, Atkinson and Lewis, 1974, Shepard, 1970, Guldmann, 1973, and Wilson and Minnotte, 1969). In general, such models overstate the cost of controlling that particular pollutant because they fail to give credit for joint reductions of other pollutants. Although there are control methods that increase the flow of some pollutants while decreasing others, it is generally the case that the emission difference terms, such as those in (1.22) and (1.23), are positive. Therefore, the level of air quality that is optimal in a one-pollutant model is likely to be less stringent than the corresponding level in a multipollutant model.

The Linear Programming Model

Let us assume that an optimal pair of air quality levels, q^1 and q^2, like those in the solution of (1.21), are known in advance. The efficient production abatement technology would be the solution of the following linear programming model:

$$\text{Maximize } Y = R - c_{1a}x_{1a} - c_{1b}x_{1b} - c_{1c}x_{1c} - d_{1d}x_{1d}$$

subject to

$$
\begin{aligned}
e^1_{1a}x_{1a} + e^1_{1b}x_{1b} + e^1_{1c}x_{1c} + e^1_{1d}x_{1d} &\leq q^1, \\
e^2_{1a}x_{1a} + e^2_{1b}x_{1b} + e^2_{1c}x_{1c} + e^2_{1d}x_{1d} &\leq q^2, \\
x_{1a} + x_{1b} + x_{1c} + x_{1d} &= \alpha R, \\
x_{1a}, x_{1b}, x_{1c}, x_{1d} &\geq 0.
\end{aligned}
\tag{1.24}
$$

In this model R is a constant; the sum of production activities for making the intermediate good is fixed at αR because of the technological assumption underlying the Pure Abatement model.

Each unit cost coefficient c_j consists of production cost and abatement cost. The latter will be called C_j. It follows that the difference

$$c_{1a} - C_{1a} = c_{1b} - C_{1b} = c_{1c} - C_{1c} = c_{1d} - C_{1d} = \sigma_1 \tag{1.25}$$

represents production cost alone. The objective function in (1.24) is equivalent to

$$
\begin{aligned}
\text{Maximize } Y = R &- (\sigma_1 + C_{1a})x_{1a} - (\sigma_1 + C_{1b})x_{1b} \\
&- (\sigma_1 + C_{1c})x_{1c} - (\sigma_1 + C_{1d})x_{1d},
\end{aligned}
\tag{1.26}
$$

which in turn is equal to

Maximize $\Upsilon = R - \sigma_1(\alpha R) - C_{1a}x_{1a}$
$$- C_{1b}x_{1b} - C_{1c}x_{1c} - C_{1d}x_{1d}. \tag{1.27}$$

Because R and $\sigma_1(\alpha R)$ are constants, maximization of Υ implies minimization of the negative quantities. Accordingly, we may substitute a new objective function

$$\text{Minimize } Z = C_{1a}x_{1a} + C_{1b}x_{1b} + C_{1c}x_{1c} + C_{1d}x_{1d}, \tag{1.28}$$

where Z is the total cost of pollution abatement.

The larger the outlay for abatement Z, the less the production of the composite good Υ, which is a residual. The true cost of cleaner air is the opportunity cost of foregone consumption of the composite good. The value of the foregone comsumption is actually greater than Z. This is illustrated in figure 1.5, which depicts the demand curve (a rectangular hyperbola) of the ith household for the composite good at a given level of income. Prior to the establishment of, say, emission standards, the price of the composite good is $\hat{p}$. After abatement the price is $\bar{p}$. The cost of abatement borne by the ith household is Z_i, which is the area of either rectangle in figure 1.5. The reduced consumption $\hat{\Upsilon}_i - \bar{\Upsilon}_i$ is undervalued at $\hat{p}$ and overvalued at $\bar{p}$. The correct value of the foregone consumption, if we may borrow the partial equilibrium concept of consumer's surplus, is Z_i plus the shaded area under the demand curve.

In the Linear Programming Model, the optimal pollutant concentrations are achieved by technological abatement alone. There are no contractions in units of output, nor substitutions in consumption, that would diminish the level of the polluting activity itself. Thus, our linear programming model has the special properties of the Pure Abatement model.

Shadow Prices and Convexity

The revised linear programming model is

$$\text{Minimize } Z = C_{1a}x_{1a} + C_{1b}x_{1b} + C_{1c}x_{1c} + C_{1d}x_{1d}$$

subject to

$$e^1_{1a}x_{1a} + e^1_{1b}x_{1b} + e^1_{1c}x_{1c} + e^1_{1d}x_{1d} \leqslant q^1,$$
$$e^2_{1a}x_{1a} + e^2_{1b}x_{1b} + e^2_{1c}x_{1c} + e^2_{1d}x_{1d} \leqslant q^2,$$
$$x_{1a} + x_{1b} + x_{1c} + x_{1d} = \alpha R, \tag{1.29}$$
$$x_{1a}, x_{1b}, x_{1c}, x_{1d} \geqslant 0.$$

The Linear Programming Model

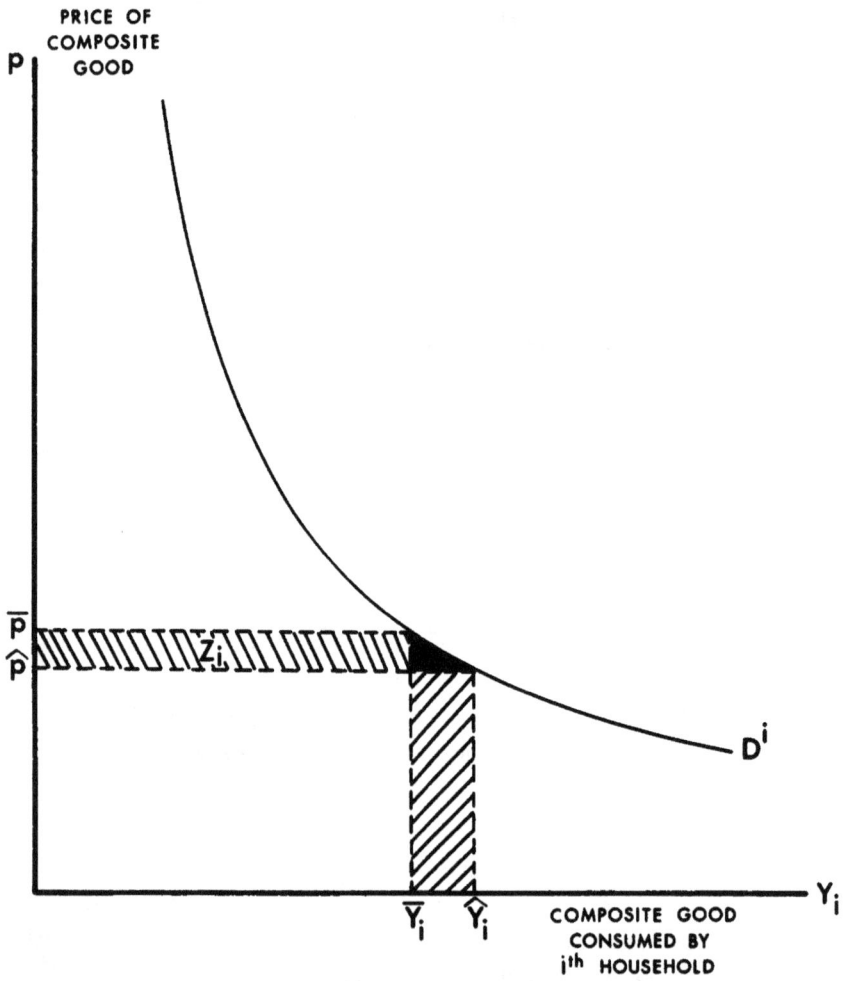

Figure 1.5
A household's demand curve for the composite good

Associated with the solution of (1.29) is a set of shadow prices $\{\pi^1, \pi^2, \pi_1\}$, each of which represents the change in $\mathcal{Z}$ when the corresponding constraint (on q^1, q^2, or αR, respectively) is increased by one unit, holding the other two constraints constant. For a solution in which, say, process 1a, 1b, and 1c are active, the vector of shadow prices $[\pi^1, \pi^2, \pi_1]$ is equal to the cost vector times the inverse of the basis matrix (see Hadley, 1962, p. 230), that is,

$$[C_{1a}\ C_{1b}\ C_{1c}] \begin{bmatrix} e^1_{1a} & e^1_{1b} & e^1_{1c} \\ e^2_{1a} & e^2_{1b} & e^2_{1c} \\ 1 & 1 & 1 \end{bmatrix}^{-1}. \tag{1.30}$$

There is a method for calculating shadow prices that is based on Cramer's Rule and is particularly useful when more than two pollutants are involved. In this method (see Dwyer, 1951, p. 138), the ith shadow price (that is, the ith entry in the row vector of shadow prices) is equal to the following ratio of determinants. The denominator of the ratio is the determinant of the basis matrix and the numerator is that same determinant with the ith row replaced by the row vector of abatement costs. For example, π^2 in (1.30) is

$$\pi^2 = \frac{\begin{vmatrix} e^1_{1a} & e^1_{1b} & e^1_{1c} \\ C_{1a} & C_{1b} & C_{1c} \\ 1 & 1 & 1 \end{vmatrix}}{\begin{vmatrix} e^1_{1a} & e^1_{1b} & 1_{1c} \\ e^2_{1a} & e^2_{1b} & e^2_{1c} \\ 1 & 1 & 1 \end{vmatrix}}. \tag{1.31}$$

Using (1.25) to express the results in terms of the c_j, it follows that

$$\pi^2 = \frac{-(c_{1c} - c_{1b})(e^1_{1a} - c^1_{1b}) + (c_{1b} - c_{1a})(e^1_{1b} - e^1_{1c})}{(e^1_{1a} - e^1_{1b})(e^2_{1b} - e^2_{1c}) - (e^1_{1b} - e^1_{1c})(e^2_{1a} - e^2_{1b})}. \tag{1.32}$$

By similar calculations,

$$\pi^1 = \frac{(c_{1c} - c_{1b})(e^2_{1a} - e^2_{1b}) - (c_{1b} - c_{1a})(e^2_{1b} - e^2_{1c})}{(e^1_{1a} - e^1_{1b})(e^2_{1b} - e^2_{1c}) - (e^1_{1b} - e^1_{1c})(e^2_{1a} - e^2_{1b})} \tag{1.33}$$

and

$$\pi_1 = c_{1a} - e^1_{1a}\pi^1 - e^2_{1a}\pi^2. \tag{1.34}$$

The meaning of π_1 is as follows. If the output of the intermediate good $s_1 = \alpha R$ is increased by one unit and the allowable flows of the two pollutants are held constant, there will be an increase in resource cost equal to c_{1a}, which is the direct labor cost of producing good-one with process 1a, plus the costs of eliminating the incremental emissions associated with one unit of activity of process 1a. These costs are the pollutant shadow prices times the incremental emissions e_{1a}^1 and e_{1a}^2. Because three processes are used, the shadow price of good-one can also be expressed in terms of the coefficients for process 1b or 1c; that is,

$$\pi^1 = c_{1b} - e_{1b}^1\pi^1 - e_{1b}^2\pi^2 = c_{1c} - e_{1c}^1\pi^1 - e_{1c}^2\pi^2. \qquad (1.35)$$

One of the conditions for a convex production-possibility frontier between Y, q^1, and q^2 is that the right-hand-side values in (1.32) and (1.33) be negative.[6] This imposes certain conditions on the values of the coefficients. Observe that the denominators in (1.32) and (1.33) are identical and can be either positive or negative. For both right-hand-side values to be negative, the following conditions must hold. Let us first assume that the six terms that comprise these ratios are all positive. If the common denominator is less than zero, it must be the case that

$$\frac{c_{1b} - c_{1a}}{e_{1a}^2 - e_{1b}^2} < \frac{c_{1c} - c_{1b}}{e_{1b}^2 - e_{1c}^2}, \qquad (1.36)$$

$$\frac{c_{1b} - c_{1a}}{e_{1a}^1 - e_{1b}^1} > \frac{c_{1c} - c_{1b}}{e_{1b}^1 - e_{1c}^1}.$$

If the common denominator is greater than zero, the inequalities in (1.36) are reversed.

If one of the six difference terms, say $(e_{1a}^2 - e_{1b}^2)$, is less than or equal to zero, it is only necessary that

$$\frac{c_{1b} - c_{1a}}{e_{1a}^1 - e_{1b}^1} < \frac{c_{1c} - c_{1b}}{e_{1b}^1 - e_{1c}^1}. \qquad (1.37)$$

If $(e_{1b}^2 - e_{1c}^2)$ alone is less than or equal to zero, it is only necessary that

$$\frac{c_{1b} - c_{1a}}{e_{1a}^1 - e_{1b}^1} < \frac{c_{1c} - c_{1b}}{e_{1b}^1 - e_{1c}^1}. \qquad (1.38)$$

If both $(e_{1a}^2 - e_{1b}^2)$ and $(e_{1b}^2 - e_{1c}^2)$ are negative, the same argument that led to condition (1.36) applies. The reader may determine the remaining conditions on the signs of the emission difference terms such that both right-hand-side ratios in (1.32) and (1.33) are negative. It is of interest that these conditions allow for production-abatement processes that increase the flow of some pollutants while decreasing that of others.

Multiple Sources of Pollution

We have assumed that all of the air pollution occurs in the production of a single intermediate good. We extend the model by assuming that there are many such intermediate goods, each of which is required by firms in some proportion α_j to their labor input. In the case of two intermediate outputs s_1 and s_2, which may be produced by a combination of two and three processes, respectively, we have

$$
\begin{aligned}
\alpha_1(c_{1a}x_{1a} + c_{1b}x_{1b} + c_{2a}x_{2a} + c_{2b}x_{2b} + c_{2c}x_{2c} + \varUpsilon) = x_{1a} + x_{1b} = s_1, \\
\alpha_2(c_{1a}x_{1a} + c_{1b}x_{1b} + c_{2a}x_{2a} + c_{2b}x_{22} + c_{2c}x_{2c} + \varUpsilon) = x_{2a} + x_{2b} + x_{2c} = s_2.
\end{aligned}
\tag{1.39}
$$

It is equivalent to view the s_i not as intermediate goods but as production activity levels that take place in the factories where final goods are manufactured. In this model there are no price-induced shifts in consumption by households that would alter the polluting activity levels s_i. This together with the assumption that the single resource is inelastically supplied fixes the levels of the outputs that are polluting in production. Furthermore, each household can be presumed to have a utility function (in terms of the pollutant levels and a composite good) that does not shift with the level of abatement outlay $\mathcal{Z}$.

These assumptions underlie the empirical model presented in chapters 2, 3, 4, and 5. However, they are not crucial to empirical analysis, and in the final chapters the model is revised to allow for changes in the quantity of resources and for shifts in relative prices with consequent feedbacks on the level of polluting activities.

APPENDIX: EQUIVALENCE OF EMISSION STANDARDS AND POLLUTION FEES AND A NUMERICAL EXAMPLE

In this appendix we demonstrate that in the Pure Abatement model, economic efficiency can be achieved by either emission standards or pollution fees. This is illustrated with a numerical example.

Conditions (1.16) and (1.17) may be satisfied in a competitive market economy by either emission standards or emission fees. If the government could predetermine the optimal level of q^* and the corresponding abatement activities such that

$$e_{1b}x_{1b}^* + e_{1c}x_{1c}^* = q^*, \qquad (1.40)$$

it could establish an emission standard of $[\gamma e_{1b} + (1 - \gamma)e_{1c}]$ units of pollution per unit of output of the intermediate good, where

$$\gamma = x_{1b}^*/(x_{1b}^* + x_{1c}^*)$$

and $\qquad\qquad (1.41)$

$$x_{1b}^* + x_{1c}^* = s_1 = \alpha R.$$

The price of the intermediate good in this market economy would be equal to its marginal cost,

$$\begin{aligned}
p_1 &= [\gamma c_{1b} + (1 - \gamma)c_{1c}] + p_1\alpha[\gamma c_{1b} + (1 - \gamma)c_{1c}] \\
&= [\gamma c_{1b} + (1 - \gamma)c_{1c}]/[1 - \alpha(\gamma c_{1b} + (1 - \gamma)c_{1c})].
\end{aligned} \qquad (1.42)$$

The prices of the final goods would be

$$\begin{aligned}
p_2 &= c_2 + p_1\alpha c_2 = c_2(1 + \alpha p_1), \\
p_3 &= c_3 + p_1\alpha c_3 = c_3(1 + \alpha p_1).
\end{aligned} \qquad (1.43)$$

The reader may confirm, using (1.12) and (1.41), that

$$p_2 y_2 + p_3 y_3 = R. \qquad (1.44)$$

Letting one unit of R represent one dollar, it follows from (1.44) that the value of final goods will be equal to consumer income. This is essential if households are to have sufficient income to purchase the entire output of goods.

The ratio, p_1/p_2, of selling prices in (1.43) is equal to c_2/c_3, and if households maximize their utility, condition (1.17) will be satisfied. To confirm that the level of pollution is optimal, the government can ascertain that the marginal benefits of abatement, which equal $-p_2(U_q^1/U_2^1 + U_q^2/U_2^2)$, must equal the marginal cost of abatement. Allowing for purchases of the intermediate good, the latter is $[(c_{1c} - c_{1b})/(e_{1b} - e_{1c})][1 + \alpha p_1]$. This equivalence satisfies condition (1.16).

If the government controls pollution by means of emission fees, the optimal fee ϕ corresponding to (1.40) would be

$$\phi = \left(\frac{c_{1c} - c_{1b}}{e_{1b} - e_{1c}}\right)(1 + \alpha p_1); \qquad (1.45)$$

and the equilibrium price of the intermediate good would be

$$p_1 = c_{1b} + \alpha c_{1b} p_1 + e_{1b} \phi. \tag{1.46}$$

It follows by simultaneous solution that in equilibrium

$$\phi = \frac{\left(\dfrac{c_{1c} - c_{1b}}{e_{1b} - e_{1c}} \right)}{1 \quad \alpha \left[c_{1b} + e_{1b} \left(\dfrac{c_{1c} - c_{1b}}{e_{1b} - e_{1c}} \right) \right]} \tag{1.47}$$

and

$$p_1 = \frac{c_{1b} + e_{1b} \left(\dfrac{c_{1c} - c_{1b}}{e_{1b} - e_{1c}} \right)}{1 - \alpha \left[c_{1b} + e_{1b} \left(\dfrac{c_{1c} - c_{1b}}{e_{1b} - e_{1c}} \right) \right]}. \tag{1.48}$$

Denoting the common denominator in (1.47) and (1.48) by Ω, it can be determined that $p_2 = c_2/\Omega$ and $p_3 = c_3/\Omega$. Assuming that producers of good-one voluntarily adopted the divisible solution $\{x_{1b}^*, x_{1c}^*\}$, conditions (1.16) and (1.17) would be satisfied in the competitive market equilibrium based on emission fees.

Although absolute prices differ if pollution is controlled by emission fees rather than by emission standards, relative prices are the same and both programs are efficient. In the absence of government control of pollution, producers of the intermediate good would use process 1a and the level of pollution would be

$$q = e_{1a} \alpha R. \tag{1.49}$$

With government control, whether by standards or fees, the level of pollution would be

$$q = [\gamma e_{1b} + (1 - \gamma) e_{1c}] \alpha R. \tag{1.50}$$

The decline in pollution is a consequence of technological abatement alone. There is no income effect on output of the polluting intermediate good, nor are there any shifts in consumption that would alter the level of pollution.

This case is illustrated with the following numerical example. Consider a simple economy consisting of two households whose utility functions are

$$
\begin{aligned}
U^1 &= 1200 \ln \Upsilon_1 + 13 \ln (5000 - q^1) + 50 \ln (6000 - q^2), \\
U^2 &= 348 \ln \Upsilon_2 + \ln (5000 - q^1) + 10 \ln (6000 - q^2),
\end{aligned} \tag{1.51}
$$

where Y_i is the quantity of output (represented by a composite good) consumed by the ith household and q^j is the level of the jth pollutant.

Pollution in this economy originates in the production of an intermediate good which is used by all firms including producers of the intermediate good, in quantities equal to 20 percent of their labor input. The total quantity of labor in this economy is 2500 units. Pollution can be reduced by using alternative processes for making the intermediate good. The production technology is described by the following equations:

$$
\begin{aligned}
1.1x_{1a} + 1.2x_{1b} + 1.4x_{1c} + 1.45x_{1d} + 1.8x_{1e} + Y &= 2500, \\
10x_{1a} + 8x_{1b} + 7x_{1c} + 7x_{1d} + 3x_{1e} &= q^1, \\
12x_{1a} + 10x_{1b} + 5x_{1c} + 3x_{1d} + 4x_{1e} &= q^2, \\
-.78x_{1a} - .76x_{1b} - .72x_{1c} - .71x_{1d} - .64x_{1e} + .2Y &= 0.
\end{aligned}
\tag{1.52}
$$

The first equation is the resource constraint, while the final equation is equivalent to

$$
\begin{aligned}
.2(1.1x_{1a} + 1.2x_{1b} + 1.4x_{1c} + 1.45x_{1d} + 1.8x_{1c} + Y) \\
= x_{1a} + x_{1b} + x_{1c} + x_{1d} + x_{1e}.
\end{aligned}
\tag{1.53}
$$

There are an infinite number of Pareto optimal allocations; one of these is

$$
\begin{aligned}
Y &= 1870, \quad Y_1 = 1000, \quad Y_2 = 870, \\
q^1 &= 4200, \quad q^2 = 4000, \\
x_{1a} &= 200, \quad x_{1b} = 100, \quad x_{1d} = 200.
\end{aligned}
\tag{1.54}
$$

It can be shown that this allocation satisfies conditions analogous to (1.19). When these three processes are nonzero, the rates of transformation in production are

$$\partial Y/\partial q^1 = -1/60 \quad \text{and} \quad \partial Y/\partial q^2 = -1/30.$$

The allocation satisfies the conditions for a Pareto optimum because

$$
U_{q^1}^1/U_Y^1 + U_{q^1}^2/U_Y^2 = \partial Y/\partial q^1,
$$
$$
\frac{-13Y_1}{(5000 - q_1)\,1200} + \frac{-Y_2}{(5000 - q^1)\,348} = -\frac{1}{60},
\tag{1.55}
$$

and

$$
U_{q^2}^1/U_Y^1 + U_{q^2}^2/U_Y^2 = \partial Y/\partial q^2,
$$
$$
\frac{-50Y_1}{(6000 - q^2)(1200)} - \frac{10Y_2}{(6000 - q^2)(348)} = -\frac{1}{30}.
\tag{1.56}
$$

This solution could be achieved by either of two governmental programs, emissions standards or emission fees. If the allowable rates of emissions for pollutants one and two were set at 8.4 and 8.0, respectively, producers of the intermediate good would meet the standard at least-cost by using 40 percent of process 1a, 20 percent of 1b, and 40 percent of 1d. The prices of the intermediate and composite goods would be 315/187 and 250/187, respectively (see equations 1.42 and 1.43 above); and the maximum producible output of the latter would be 2500/(250/187), or 1870 units.

Alternatively, emission fees of 1/40 per unit of pollutant-one and 1/20 per unit of pollutant-two would result in market prices of 2.5 and 1.5 for the intermediate and composite goods respectively.[7] These fees would prompt producers of the intermediate good to use some combination of processes 1a, 1b, and 1d. Assuming, for convenience, that the 40 percent, 20 percent, 40 percent combination were chosen, the pollution levels would be $q^1 = 4200$ and $q^2 = 4000$. The fee revenue to the government, which would equal 305, could be transferred to the households so that disposable income would be 2805. The equilibrium output of the composite good would therefore be 2805/1.5, or 1870.

A simplified version of the model is one in which the Pareto optimal levels of pollution are given in advance and a least-cost set of abatement processes determined. Assuming that the pure production cost of the intermediate good is 1.0, the optimal combination of processes is the solution of the following linear programming problem:

Minimize

$$Z = .1x_{1a} + .2x_{1b} + .4x_{1c} + .45x_{1d} + .8x_{1e}$$

subject to

$$
\begin{aligned}
10x_{1a} + 8x_{1b} + 7x_{1c} + 7x_{1d} + 3x_{1e} &\leq 4200, \\
12x_{1a} + 10x_{1b} + 5x_{1c} + 3x_{1d} + 4x_{1e} &\leq 4000, \\
x_{1a} + x_{1b} + x_{1c} + x_{1d} + x_{1e} &= 500, \\
x_{1a}, x_{1b}, x_{1c}, x_{1d}, x_{1e} &\geq 0.
\end{aligned}
\tag{1.57}
$$

The objective function of (1.57) is the total cost of abatement, while the equality constraint is derived from (1.52).

The five processes in (1.57) define a production-possibility frontier between q^1, q^2, and Z. This frontier is illustrated in figure 1.6. The facets of

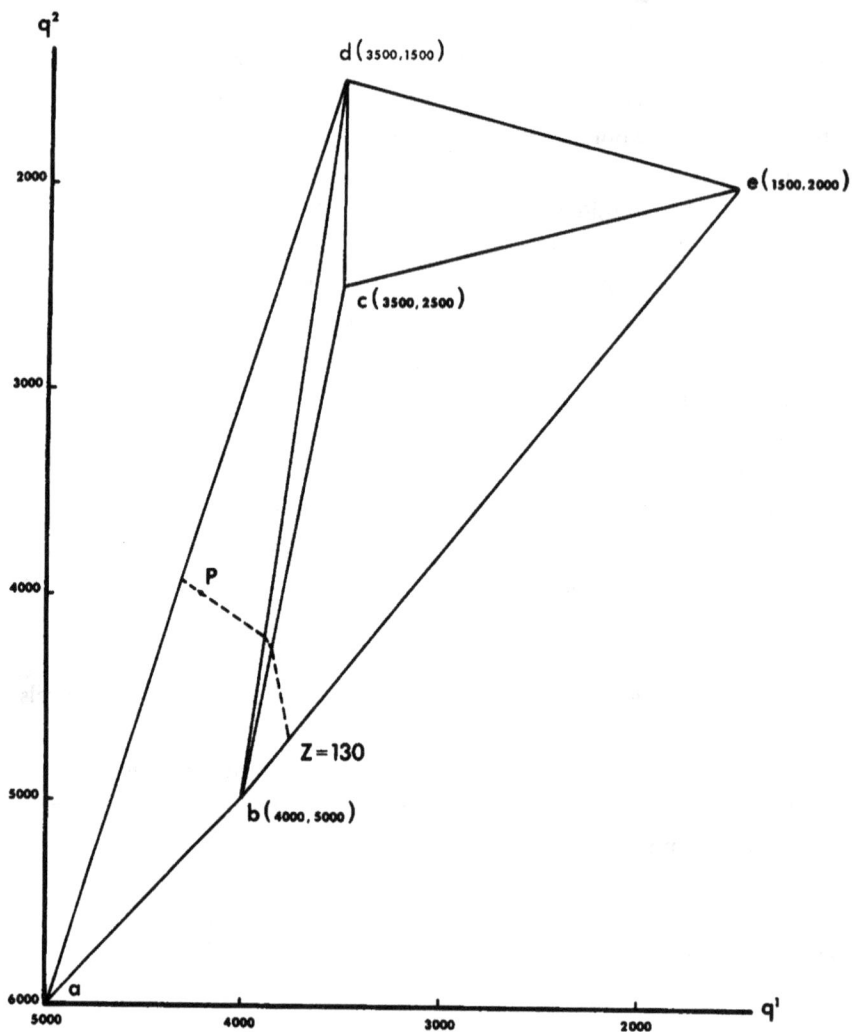

Figure 1.6
Production-possibility frontier for the numerical example

this frontier are projected on to the $q^1 q^2$ plane. Because there are five pro-cesses, there can be no more than four facets containing technically ef-ficient combinations of outputs. There are three other combinations of processes, (1a, 1b, 1c), (1a, 1d, 1e), and (1b, 1d, 1e), that generate facets with negative pollutant shadow prices; however, these facets lie behind the ones shown in figure 1.6. The pollutant shadow prices for the respective facets are as follows:

facet abd : $\pi^1 = -.01667$, $\pi^2 = -.03333$;
facet bcd : $\pi^1 = -.075$, $\pi^2 = -.025$;
facet cde : $\pi^1 = -.09375$, $\pi^2 = -.025$;
facet bce : $\pi^1 = -.09474$, $\pi^2 = -.02105$.

$$(1.58)$$

The optimal solution to (1.58) is denoted in figure 1.6 by the point P on the dotted frontier $Z = 130$. Frontiers for higher values of Z are northeast of the one illustrated.

Note that in moving along an isoquant such as $Z = 130$, the shadow price of one pollutant decreases while that of the other increases (or, in one case, stays the same). This is a condition of convexity of the production-possibility frontier.

This numerical example illustrates that in the case of the Pure Abatement model, both emission standards and pollution fees can promote economic efficiency.

2

In the preceding chapter a model was developed in which pollution occurs in the production of intermediate goods or activities. By defining a technology in which the quantities of such intermediate activity levels are proportional to labor input, we provided a theoretical basis for a linear programming model in which each of the polluting activity levels is constant.

In this chapter the Linear Programming Model is generalized in matrix notation. The numerical data for implementing the model are presented. These data are used to estimate the total emission flows in the St. Louis airshed in 1975 in the absence of a regulatory program. The allowable flows are deducted from the projected flows to determine the quantities of emission abatement that are necessary to achieve the desired level of air quality.

Format of the Linear Programming Model

The Linear Programming Model (1.29) can be expanded to include m intermediate outputs that are polluting in production. This set of outputs, $\{s_1, s_2, s_3, \ldots, s_m\}$, is represented by an $m \times 1$ column vector $\mathbf{s}$. We shall refer to these outputs as pollution source levels.

Based on the theory in the preceding chapter, the elements of the $\mathbf{s}$ vector are assumed to be constants. In keeping with the theoretical model, the s_i should be restricted to intermediate outputs produced by users (such as British thermal units of heat or tons of waste incinerated) as well as intermediate outputs produced for sale to other producers (such as kilowatt hours of electricity or tons of iron). In the empirical model some of the s_i are final consumption outputs; nevertheless, we retain the assumption that all of the source levels are constant. These source magnitudes are measured in units such as tons of coal burned or gallons of gasoline consumed.

Corresponding to each pollution-related output, say s_2, there is an activity variable x_{2a} that represents the base-year level of abatement. In some cases this is the only activity variable corresponding to that particular source of pollution, indicating that there are no abatement alternatives. When a set of abatement alternatives do exist for s_2, the corresponding activity levels are $x_{2b'}, x_{2c'}, x_{2d'}$, etc. The entire set of activity levels is an $n \times 1$ vector, $\mathbf{x}$. We shall refer to the typical element of that vector as simply x_j. The sum of activity levels for any polluting source i equals the constant s_i. Thus we may have

$$x_{1a} + x_{1b} = s_1,$$
$$x_{2a} + x_{2b} + x_{2c} = s_2, \tag{2.1}$$
$$x_{3a} = s_3.$$

This may be rewritten as follows:

$$1x_{1a} + 1x_{1b} + 0x_{2a} + 0x_{2b} + 0x_{2c} + 0x_{3a} = s_1,$$
$$0x_{1a} + 0x_{1b} + 1x_{2a} + 1x_{2b} + 1x_{2c} + 0x_{3a} = s_2, \tag{2.2}$$
$$0x_{1a} + 0x_{1b} + 0x_{2a} + 0x_{2b} + 0x_{2c} + 1x_{3a} = s_3.$$

In matrix notation (2.2) is equivalent to

$$\mathbf{ux} = \mathbf{s}, \tag{2.3}$$

where $\mathbf{u}$ is the distributive matrix,

$$\mathbf{u} = \begin{bmatrix} 1 & 1 & 0 & 0 & 0 & 0 \\ 0 & 0 & 1 & 1 & 1 & 0 \\ 0 & 0 & 0 & 0 & 0 & 1 \end{bmatrix}. \tag{2.4}$$

In the Linear Programming Model, $\mathbf{u}$ is an $m \times n$ matrix whose typical element, u_{ij}, is unity when the jth activity is defined for the ith pollution source and is zero otherwise. Although the elements of the $\mathbf{s}$ vector are numbered from 1 to m, the elements of the $\mathbf{x}$ vector are alphanumerically ordered and are not, therefore, actually numbered from 1 to n. For convenience, however, we shall refer to the final element of the $\mathbf{x}$ vector as x_n.

The objective function of the Linear Programming Model is the summation of abatement outlays

$$\text{Minimize } Z = C_{1a}x_{1a} + C_{1b}x_{1b} + \quad + C_j x_j + \cdots + C_n x_n, \tag{2.5}$$

which in matrix notation is

$$\text{Minimize } \mathbf{Z} = \mathbf{Cx}, \tag{2.6}$$

where $\mathbf{C}$ is the $1 \times n$ row vector of abatement costs. If an element x_j represents nonabatement, then C_j will be zero.

Associated with the activity variable x_j is a $p \times 1$ column vector whose element e_j^i denotes emissions of the ith pollutant per unit of the jth activity. Thus the total emission flows f^i in the economy are

$$e_{1a}^1 x_{1a} + e_{1b}^1 x_{1b} + \cdots + e_j^1 x_j + \cdots + e_n^1 x_n = f^1,$$
$$e_{1a}^p x_{1a} + e_{1b}^p x_{1b} + \cdots + e_j^p x_j + \cdots + e_n^p x_n = f^p, \tag{2.7}$$

Data for the Linear Programming Model

or in matrix notation,

ex = f, (2.8)

where **e** is a $p \times n$ matrix of emission factors and **f** is a $p \times 1$ vector of emission flows.[1] In chapters 2, 3, and 4, we shall express air quality constraints in terms of annual emission flows f^i.

The first model, which is called Model I, is

Minimize **Z = Cx**
subject to **ux = s,**
$\qquad$ **ex ⩽ f,** (2.9)
$\qquad$ **x ⩾ 0.**

The Data

The St. Louis airshed, which is illustrated in figure 2.1, includes seven political subdivisions in two states. In the Missouri portion there are the City of St. Louis and Jefferson, St. Charles, and St. Louis counties. In Illinois there are Madison, St. Clair, and Monroe counties. These comprise an area of 3600 square miles and a population of 2.5 million people. Pollution control in the region is fragmented into at least five major city, county, state, and federal agencies. In this model, it is assumed that there are specific allowable annual flows of carbon monoxide, hydrocarbons, nitrogen oxides, sulfur dioxide, particulates, and benzo(a)pyrene, and that a uniform set of pollution control methods, which achieves the allowable flows at the lowest total cost of abatement, is desired.

The data for implementing Linear Programming Model I for the year 1975 are contained in table 2.1. These data are adapted from Kohn (1969a). Columns 1 and 2 list and define 250 activity variables, which make up the **x** vector. Activities 1a, 1b, 1c, and 1d are defined for pollution source 1; activities 24a, 24b, and 24c for pollution source 24, etc.[2] For a number of pollution sources there are no alternative abatement activities, indicating either that none would be available (as in the case of sources 6, 7, 8, 9, 10, 50, and others) or that the source was already controlled to the maximum feasible level (as in the case of sources 45, 46, 47, 48, 61, 95, and others). The final activity is numbered 97b, indicating that there would be 97 different sources of air pollution in the St. Louis airshed in the year 1975.

The estimated magnitude of the pollution sources are the elements of a (97×1) vector, **s**. They are listed in column 10 of table 2.1 and are

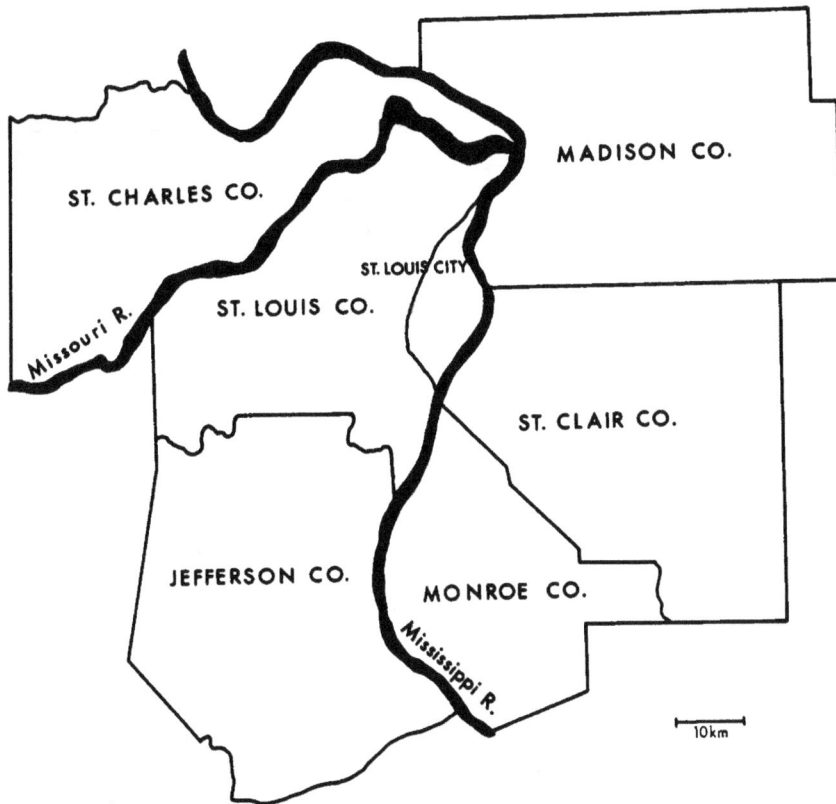

Figure 2.1
Map of the St. Louis airshed

Table 2.1

Elements of the matrices and vectors in the Linear Programming Model for air pollution control in the St. Louis airshed in 1975

(1)	(2)	(3)	(4)	(5)
Variable number j	Activity unit and description	Cost of abatement (dollars) C_j	Carbon monoxide (pounds) e_j^k	Hydro-carbons (pounds) e_j^k
1a	Thousand gallons of gasoline burned in 1967 and earlier model vehicles	0	2,910.	616.
1b	As 1a, with retrofit crankcase and exhaust controls	26.30	1,339.	265.
1c	As 1a, with retrofit device to recirculate nitrogen oxides	50.00	2,910.	616.
1d	Combination 1b and 1c	76.30	1,339.	265.
2a	Thousand gallons of gasoline burned in 1968 and 1969 model vehicles	0	2,910.	616.
	As 2a, with crankcase and exhaust controls as original equipment	19.50	1,339.	265.
	As 2a, with retrofit device to recirculate nitrogen oxides	50.00	2,910.	616.
	Combination 2b and 2c	69.50	1,339.	265.
	Thousand gallons of gasoline burned in 1970–1975 model vehicles	0	2,910.	616.
	As 3a, with device to recirculate nitrogen oxides as original equipment	10.25	2,910.	616.
	As 3b, with crankcase and exhaust controls as original equipment	28.71	931.2	212.52
	As 3c, with evaporative loss control device	30.76	931.2	129.72
	Thousand gallons of number two diesel fuel burned in buses	0	60.	180.
	As 4a, except number one diesel fuel	14.36	62.4	187.2
	Thousand gallons of number two diesel fuel burned by trucks, railroads, and river vessels	0	60.	180.

(6)	(7)	(8)	(9)	(10)	(11)
Nitrogen oxides (pounds) e_j^*	Sulfur dioxide (pounds) e_j^*	Particulate matter (pounds) e_j^*	Benzo(a)- pyrene (thousand micrograms) e_j^*	Source magnitude s_i	Optimal activity level x_j^*
113.	9.	11.	270.	330,700	330,700
124.3	9.	11.	270.	330,700	0
22.6	9.	11.	270.	330,700	0
33.9	9.	11.	270.	330,700	0
113.	9.	11.	270.	180,800	180,800
124.3		11.	270.	180,800	0
22.6		11.	270.	180,800	0
33.9		11.	270.	180,800	0
113.		11.	270.	625,500	0
22.6		11.	270.	625,500	79,862
22.6		11.	270.	625,500	519,753
22.6		11.	270.	625,500	25,885
222.		110.	400.	7,252	7,252
230.9		53.5	416.	7,252	0
222.		110.	400.	69,350	69,350

Data for the Linear Programming Model

Table 2.1 (Continued)

(1)	(2)	(3)	(4)	(5)
Variable number j	Activity unit and description	Cost of abatement (dollars) C_j	Carbon monoxide (pounds) e_j^k	Hydro-carbons (pounds) e_j^k
5b	As 5a, except number one diesel fuel	28.72	62.4	187.2
6a	Jet aircraft flights (one flight equals a take-off and a landing)	0	35.	10.
7a	Two-engine turboprop flights	0	2.02	.27
8a	Four-engine turboprop flights	0	8.71	1.18
9a	Two-engine piston aircraft flights	0	134.	25.
10a	Four-engine piston aircraft flights	0	326.	60.
11a	Thousand gallons of number six fuel oil burned by industry	0	.04	3.2
11b	As 11a, except low sulfur content oil	2.50	.04	3.2
11c	As 11a, except number two fuel oil	33.50	.043	3.408
12a	Thousand gallons of number five fuel oil burned by industry	0	.04	3.2
12b	As 12a, except low sulfur content	5.00	.04	3.2
13a	Thousand gallons of residual fuel oil burned by electric utilities	0	.04	3.2
14a	Thousand gallons of distillate fuel oil burned by industry	0	2.	2.
15a	Thousand gallons of distillate fuel oil burned in residences	0	2.	2.
16a	Thousand gallons of distillate fuel oil burned by commercial and institutional users	0	2.	2.
17a	Tons of coal burned by industry in unequipped underfeed stokers	0	3.	1.
17b	As 17a, with 85 percent efficient mechanical collectors	1.62	3.	1.
17c	As 17a, except low sulfur coal	2.41	3.	1.
17d	Combination of 17b and 17c	4.03	3.	1.

(6)	(7)	(8)	(9)	(10)	(11)
Nitrogen oxides (pounds) e_j^t	Sulfur dioxide (pounds) e_j^t	Particulate matter (pounds) e_j^t	Benzo(a)-pyrene (thousand micrograms) e_j^t	Source magnitude s_i	Optimal activity level x_j^*
230.9	16.	53.5	416.	69,350	0
23.	1.8	34.	0	136,500	136,500
1.13	2.0	.59	0	14,700	14,700
4.86	2.0	2.54	0	2,100	2,100
6.3	.02	.6	0	52,500	52,500
15.4	.02	1.4	0	4,200	4,200
104.	277.356	8.	5.	39,700	0
104.	159.4	8.	5.	39,700	39,700
110.76	44.815	8.52	5.	39,700	0
104.	232.724	8.	5.	158,800	158,800
104.	159.4	8.	5.	158,800	0
104.	277.356	8.	5.	0	0
72.	41.98	12.	40.	11,540	11,540
72.	41.98	12.	40.	150,100	150,100
72.	41.98	12.	40.	8,017	8,017
20.	117.8	50.	100.	580	0
20.	117.8	7.5	100.	580	0
20.	60.8	50.	100.	580	0
20.	60.8	7.5	100.	580	0

Data for the Linear Programming Model

Table 2.1 (Continued)

(1)	(2)	(3)	(4)	(5)
Variable number j	Activity unit and description	Cost of abatement (dollars) C_j	Carbon monoxide (pounds) e_j^*	Hydro-carbons (pounds) e_j^*
17e	As 17a, except replaced by natural gas	6.16	.007	7.413
18a	Tons of coal burned by industry in underfeed stokers equipped with some pollution controls	.63	3.	1.
18b	As 18a, with 85 percent efficient mechanical collectors	1.85	3.	1.
18c	As 18a, except low sulfur coal	3.04	3.	1.
18d	Combination of 18b and 18c	4.26	3.	1.
18e	As 18a, except replaced by natural gas	6.70	.007	7.413
19a	Tons of coal burned by industry in unequipped chain grate strokers	0	3.	1.
19b	As 19a, with 95 percent efficient mechanical collectors	.73	3.	1.
19c	As 19a, except low sulfur coal	1.25	3.	1.
19d	Combination of 19b and 19c	1.98	3.	1.
19e	As 19a, except replaced by natural gas	8.34	.008	9.216
20a	Tons of coal burned by industry in chain grate stokers equipped with settling chambers or water sprays	.28	3.	1.
20b	As 20a, with 95 percent efficient mechanical collectors	.83	3.	1.
20c	As 20a, except low sulfur ccal	1.53	3.	1.
20d	Combination of 20b and 20c	2.08	3.	1.
20e	As 20a, except replaced by natural gas	8.54	.008	9.216
21a	Tons of coal burned by industry in chain grate stokers equipped with 80 percent efficient mechanical collectors	.45	3.	1.
21b.	As 21a, with 95 percent efficient mechanical collector	.81	3.	1.

(6)	(7)	(8)	(9)	(10)	(11)
			Benzo(a)-		
Nitrogen	Sulfur	Particulate	pyrene		
oxides	dioxide	matter	(thousand	Source	Optimal
(pounds)	(pounds)	(pounds)	micrograms)	magnitude	activity level
e_j^t	e_j^t	e_j^t	e_j^t	s_i	x_j^*
3.638	.007	.306	.34	580	580
20.	117.8	31.25	100.	2,400	0
20.	117.8	7.5	100.	2,400	0
20.	60.8	31.25	100.	2,400	0
20.	60.8	7.5	100.	2,400	0
3.638	.007	.306	.34	2,400	2,400
20.	117.8	50.	100.	173,000	0
20.	117.8	2.5	100.	173,000	0
20.	60.8	50.	100.	173,000	0
20.	60.8	2.5	100.	173,000	66,970
4.518	.008	.38	.421	173,000	106,030
20.	117.8	31.25	100.	358,800	0
20.	117.8	2.50	100.	358,800	142,000
20.	60.8	31.25	100.	358,800	0
20.	60.8	2.50	100.	358,800	216,800
4.518	.008	.38	.421	358,800	0
20.	117.8	10.	100.	97,600	0
20.	117.8	2.5	100.	97,600	0

Data for the Linear Programming Model

Table 2.1 (Continued)

(1) Variable number j	(2) Activity unit and description	(3) Cost of abatement (dollars) C_j	(4) Carbon monoxide (pounds) e_j^1	(5) Hydro-carbons (pounds) e_j^2
21c	As 21a, except low sulfur coal	1.70	3.	1.
21d	Combination of 21b and 21c	2.06	3.	1.
21e	As 21a, except replaced by natural gas	8.62	.008	9.216
22a	Tons of coal burned by industry in pulverized coal units equipped with mechanical collectors	.43	3.	1.
22b	As 22a, with electrostatic precipitators	1.19	3.	1.
22c	As 22a, except low sulfur coal	1.68	3.	1.
22d	Combination of 22b and 22c	2.44	3.	1.
22e	As 22a, except replaced by natural gas	10.53	.009	10.163
23a	Tons of coal burned by industry in pulverized coal units equipped with 90 percent efficient electrostatic precipitators	.90	3.	1.
23b	As 23a with 98 percent efficient electrostatic precipitators	1.40	3.	1.
23c	As 23a, except low sulfur coal	2.15	3.	1.
23d	Combination of 23b and 23c	2.65	3.	1.
23e	As 23a, except replaced by natural gas	10.82	.009	10.163
24a	Tons of pulverized coal burned in largest single industrial boiler in the airshed	.90	3.	1.
24b	As 24a, with stack gas desulfurization process	4.17	3.	1.
24c	As 24a, except replaced by natural gas	10.82	.009	10.163
25a	Tons of pulverized coal, which can feasibly be replaced by coke oven gas, in an industrial boiler	.90	3.	1.

(6) Nitrogen oxides (pounds) e_j^t	(7) Sulfur dioxide (pounds) e_j^t	(8) Particulate matter (pounds) e_j^t	(9) Benzo(a)- pyrene (thousand micrograms) e_j^t	(10) Source magnitude s_i	(11) Optimal activity level x_j^*
20.	60.8	10.	100.	97,600	0
20.	60.8	2.5	100.	97,600	97,600
4.518	.008	.38	.421	97,600	0
20.	117.8	48.	.6	114,000	0
20.	117.8	3.2	.6	114,000	114,000
20.	60.8	48.	.6	114,000	0
20.	60.8	18.24	.6	114,000	0
4.988	.009	.42	.466	114,000	0
20.	117.8	16.	.6	444,000	0
20.	117.8	3.2	.6	444,000	444,000
20.	60.8	30.4	.6	444,000	0
20.	60.8	18.24	.6	444,000	0
4.988	.009	.42	.466	444,000	0
20.	117.8	16.	.6	110,000	0
16.	11.78	.08	.6	110,000	110,000
4.988	.009	.42	.466	110,000	0
20.	117.8	16.	.6	29,000	0

Data for the Linear Programming Model

Table 2.1 (Continued)

(1)	(2)	(3)	(4)	(5)
Variable number j	Activity unit and description	Cost of abatement (dollars) C_j	Carbon monoxide (pounds) e_j^k	Hydro-carbons (pounds) e_j^k
25b	As 25a, except replaced by coke oven gas	.48	.005	0
26a	Tons of coal burned by industry in spreader stokers with ash reinjection and 75 percent efficient mechanical collectors	.45	3.	1.
26b	As 26a, with 92.5 percent efficient mechanical collectors	.77	3.	1.
26c	As 26a, except low sulfur coal	1.70	3.	1.
26d	Combination of 26b and 26c	2.02	3.	1.
26e	As 26a, except replaced by natural gas	7.18	.008	8.967
27a	Tons of coal burned by industry in spreader stokers with ash reinjection and 90 percent efficient mechanical collectors	.56	3.	1.
27b	As 27a, with 92.5 percent efficient mechancial collectors	.77	3.	1.
27c	As 27a, except low sulfur coal	1.81	3.	1.
27d	Combination of 27b and 27c	2.02	3.	1.
27e	As 27a, except converted to natural gas	7.03	.008	8.967
28a	Tons of coal burned by industry in spreader stokers without ash reinjection and unequipped	0	3.	1.
28b	As 28a, with 92.5 percent efficient mechanical collectors	.64	3.	1.
28c	As 28a, except low sulfur coal	1.25	3.	1.
28d	Combination of 28b and 28c	1.89	3.	1.
28e	As 28a, except converted to natural gas	6.48	.008	8.608

(6) Nitrogen oxides (pounds) e_j^t	(7) Sulfur dioxide (pounds) e_j^t	(8) Particulate matter (pounds) e_j^t	(9) Benzo(a)-pyrene (thousand micrograms) e_j^t	(10) Source magnitude s_i	(11) Optimal activity level x_j^*
2.675	95.	.225	.25	29,000	29,000
20.	117.8	50.	.7	35,000	0
20.	117.8	15.	.7	35,000	0
20.	60.8	50.	.7	35,000	0
20.	60.8	15.	.7	35,000	0
4.401	.008	.37	.411	35,000	35,000
20.	117.8	20.	.7	63,500	0
20.	117.8	15.	.7	63,500	0
20.	60.8	20.	.7	63,500	0
20.	60.8	15.	.7	63,500	0
4.401	.008	.37	.411	63,500	63,500
20.	117.8	130.	.7	8,000	0
20.	117.8	9.75	.7	8,000	0
20.	60.8	130.	.7	8,000	0
20.	60.8	9.75	.7	8,000	0
4.225	.008	.356	.4	8,000	8,000

Data for the Linear Programming Model

Table 2.1 (Continued)

(1) Variable number j	(2) Activity unit and description	(3) Cost of abatement (dollars) C_j	(4) Carbon monoxide (pounds) e_j^b	(5) Hydro-carbons (pounds) e_j^b
29a	Tons of coal burned by industry in spreader stokers without ash reinjection, equipped with 30 percent efficient mechanical collectors	.28	3.	1.
29b	As 29a, with 92.5 percent efficient mechanical collectors	.76	3.	1.
29c	As 29a, except low sulfur coal	**1.53**	3.	1.
29d	Combination of 29b and 29c	**2.01**	3.	1.
29e	**As 29a, except converted to natural gas**	**6.67**	.008	8.608
30a	Tons of coal burned by industry in spreader stokers without ash reinjection with 75 percent efficient mechanical collectors	.45	3.	1.
30b	As 30a, with 92.5 percent efficient mechanical collectors	.77	3.	1.
30c	As 30a, except low sulfur coal	1.70	3.	1.
30d	Combination of 30b and 30c	2.02	3.	1.
30e	As 30a, except replaced by natural gas	6.75	.008	8.608
31a	Tons of coal burned by industry in spreader stokers without ash reinjection with 90 percent efficient mechanical collectors	.56	3.	1.
31b	As 31a, with 92.5 percent efficient mechanical collectors	.77	3.	1.
31c	As 31a, except low sulfur coal	1.81	3.	1.
31d	Combination of 31b and 31c	2.02	3.	1.
31e	As 31a, except replaced by natural gas	6.69	.008	8.608
32a	Tons of coal burned by industry in hand fired furnaces	0	3.	1.
32b	As 32a, except low sulfur coal	2.41	3.	1.

(6)	(7)	(8)	(9)	(10)	(11)
Nitrogen oxides (pounds) e_j^t	Sulfur dioxide (pounds) e_j^t	Particulate matter (pounds) e_j^t	Benzo(a)-pyrene (thousand micrograms) e_j^t	Source magnitude s_t	Optimal activity level x_j^*
20.	117.8	91.	.7	17,000	0
20.	117.8	9.75	.7	17,000	0
20.	60.8	91.	.7	17,000	0
20.	60.8	9.75	.7	17,000	0
4.225	.008	.356	.4	17,000	17,000
20.	117.8	32.5	.7	28,000	0
20.	117.8	9.75	.7	28,000	0
20.	60.8	32.5	.7	28,000	0
20.	60.8	9.75	.7	28,000	0
4.225	.008	.356	.4	28,000	28,000
20.	117.8	13.	.7	15,000	0
20.	117.8	9.75	.7	15,000	0
20.	60.8	13.	.7	15,000	0
20.	60.8	9.75	.7	15,000	0
4.225	.008	.356	.4	15,000	15,000
20.	117.8	20.	12,000.	1,120	0
20.	60.8	20.	12,000.	1,120	0

Table 2.1 (Continued)

(1)	(2)	(3)	(4)	(5)
Variable number j	Activity unit and description	Cost of abatement (dollars) C_j	Carbon monoxide (pounds) e_j^*	Hydro-carbons (pounds) e_j^*
32c	As 32a, except converted to natural gas	2.47	.005	5.978
33a	Tons of coal burned in residential stokers	0	50.	10.
33b	As 33a, except converted to natural gas	3.06	.007	7.413
34a	Tons of coal burned in residential hand-fired furnaces	0	50.	10.
34b	As 34a, except converted to natural gas	2.47	.006	7.413
35a	Tons of coal burned by commercial and institutional users in chain grate stokers with some controls	.63	32.	6.5
35b	As 35a, with improved mechanical collectors	1.85	32.	6.5
35c	As 35a, except low sulfur coal	3.04	32.	6.5
35d	Combination of 35b and 35c	4.26	32.	6.5
35e	As 35a, except replaced by natural gas	11.76	.008	9.207
36a	Tons of coal burned by commercial and institutional users in un-equipped chain grate stokers	0	32.	6.5
36b	As 36a, with 85 percent efficient mechanical collectors	1.62	32.	6.5
36c	As 36a, except low sulfur coal	2.41	32.	6.5
36d	Combination of 36b and 36c	4.03	32.	6.5
36e	As 36a, except converted to natural gas	11.21	.008	9.206
37a	Tons of coal burned by commercial and institutional users in miscellaneous equipment	0	32.	6.5
37b	As 37a, with improved mechanical collectors	1.62	32.	6.5
37c	As 37a, except low sulfur coal	2.41	32.	6.5

(6) Nitrogen oxides (pounds) e_j^t	(7) Sulfur dioxide (pounds) e_j^t	(8) Particulate matter (pounds) e_j^t	(9) Benzo(a)-pyrene (thousand micrograms) e_j^t	(10) Source magnitude s_i	(11) Optimal activity level x_i^*
1.59	.005	.26	1.781	1,120	1,120
8.	117.8	50.	100.	428,000	0
1.972	.007	.322	2.208	428,000	428,000
8.	117.8	20.	12.000	32,000	0
3.978	.005	.26	1.781	32,000	32,000
8.	117.8	25.	100.	24,000	24,000
8.	117.8	12.5	100.	24,000	0
8.	60.8	25.	100.	24,000	0
8.	60.8	12.5	100.	24,000	0
2.449	.008	.4	2.742	24,000	0
8.	117.8	50.	100.	60,000	0
8.	117.8	7.5	100.	60,000	60,000
8.	60.8	50.	100.	60,000	0
8.	60.8	7.5	100.	60,000	0
2.449	.008	.4	2.742	60,000	0
8.	117.8	50.	100.	44,000	0
8.	117.8	7.5	100.	44,000	0
8.	60.8	50.	100.	44,000	0

Data for the Linear Programming Model

Table 2.1 (Continued)

(1) Variable number j	(2) Activity unit and description	(3) Cost of abatement (dollars) C_j	(4) Carbon monoxide (pounds) e_j^*	(5) Hydro-carbons (pounds) e_j^*
37d	Combination of 37b and 37c	4.03	32.	6.5
37e	As 37a, except converted to natural gas	6.16	.007	7.41
38a	Tons of coal burned in the Wood River power plant	.16	.5	.2
38b	As 38a, with electrostatic precipitators	.79	.5	.2
38c	As 38a, except low sulfur coal	1.21	.5	.2
38d	As 38a, with stack gas desul-furization process	1.43	.5	.2
38e	As 38a, except replaced by natural gas	9.36	0	10.2
39a	Tons of coal burned in the Ashley power plant	.30	.5	.2
39b	As 39a, with upgraded electro-static precepitator	.80	.5	.2
39c	As 39a, except low sulfur coal	1.35	.5	.2
39d	Combination of 39b and 39c	1.85	.5	.2
40a	Tons of coal burned in the Venice power plant	.35	.5	.2
40b	As 40a, except low sulfur coal	1.40	.5	.2
40c	As 40a, with new electrostatic precipitators	6.40	.5	.2
40d	Combination of 40b and 40c	7.45	.5	.2
40e	As 40a, except replaced by natural gas	9.37	0	9.35
41a	Tons of coal burned in the Meramec power plant	.50	.5	.2
41b	As 41a, except low sulfur coal	1.55	.5	.2
41c	As 41a, with stack gas desul-furization process	1.70	.5	.2

(6)	(7)	(8)	(9) Benzo(a)-pyrene	(10)	(11)
Nitrogen oxides (pounds) e_j^t	Sulfur dioxide (pounds) e_j^t	Particulate matter (pounds) e_j^t	(thousand micrograms) e_j^t	Source magnitude s_i	Optimal activity level x_j^*
8.		7.5	100.	44,000	
1.97		.322		44,000	
20.		24.		1,000,000	
20.		3.2		1,000,000	
20.		24.		1,000,000	
16.		0		1,000,000	
9.1		.35		1,000,000	
20.		16.		200,000	
20.		3.2		200,000	
20.		30.4		200,000	
20.		18.24		200,000	
20.		12.		132,000	
20.	60.8	26.24		132,000	
20.	117.8	3.2		132,000	
20.	60.8	18.25		132,000	
9.1	.009	.349		132,000	
20.	117.8	3.2		730,000	
20.	60.8	18.2		730,000	
16.	11.78	.8		730,000	

Data for the Linear Programming Model

Table 2.1 (Continued)

(1)	(2)	(3)	(4)	(5)
Variable number j	Activity unit and description	Cost of abatement (dollars) C_j	Carbon monoxide (pounds) e_j^*	Hydro-carbons (pounds) e_j^*
42a	Tons of coal burned in the Sioux power plant	.50	.295	.118
42b	As 42a, except low sulfur coal	1.55	.295	.118
42c	As 42a, with stack gas desul-furization process	1.77	.295	.118
43a	Tons of coal burned in Labadie power plant in units under construction in 1969	.55	.25	.1
43b	As 43a, with stack gas desul-furization process	1.13	.25	.1
44a	Tons of coal burned in Labadie power plant in units constructed after 1970	.55	.25	.1
44b	As 44a, except nuclear units or plant located at a coal mine mouth outside of the airshed	1.55	0	0
45a	Million cubic feet of natural gas burned in power plants	0	0	436.
46a	Million cubic feet of natural gas burned by industry	0	.4	436.
47a	Million cubic feet of natural gas burned in residences	0	.4	436.
48a	Million cubic feet of natural gas burned by commerical and institutional users	0	.4	436.
49a	Million cubic feet of untreated by-product gas burned in refineries	0	0	26.
49b	As 49a, except gas alkaline treated	67.00	0	26.
50a	Million cubic feet of alkaline treated by-product gas burned in refineries	67.00	0	26.
51a	Million cubic feet of coke oven gas burned by coke producers	0	.1	0
52a	Tons of municipal refuse burned in city incinerators	0	.3	.35

(6)	(7)	(8)	(9)	(10)	(11)
Nitrogen oxides (pounds) e_j^t	Sulfur dioxide (pounds) e_j^t	Particulate matter (pounds) e_j^t	Benzo(a)-pyrene (thousand micrograms) e_i^t	Source magnitude s_i	Optimal activity level x_i^*
11.8	62.776	.236	3.54	3,000,160	0
11.8	35.876	1.336	3.54	3,000,160	0
9.44	6.276	0	3.54	3,000,160	3,000,160
10.	58.9	.4	.3	3,430,000	0
8.	5.89	0	.3	3,430,000	3,430,000
10.	58.9	.4	.3	2,070,000	0
0	0	0	0	2,070,000	2,070,000
390.	.4	15.	0	10,500	10,500
214.	.4	18.	20.	123,000	123,000
116.	.4	19.	130.	112,500	112,500
116.	.4	19.	130.	38,000	38,000
230.	950.	20.	0	7,262	7,262
230.	190.	20.	0	7,262	0
230.	133.	20.	0	26,176	26,176
53.5	1,900.	4.5	5.	17,974	17,974
2.4	1.8	17.	6.2	357,000	357,000

Table 2.1 (Continued)

(1)	(2)	(3)	(4)	(5)
Variable number j	Activity unit and description	Cost of abatement (dollars) C_j	Carbon monoxide (pounds) e_j^*	Hydro-carbons (pounds) e_j^*
52b	As 52a, with wet scrubbers	.87	.3	.35
53a	Tons of industrial waste burned in multiple chamber incinerators	.00	.5	.25
53b	As 53a, with wet scrubbers	1.00	.5	.25
54a	Tons of residential refuse burned in flue-fed incinerators	.00	27.	2.
54b	As 54a, except disposal in sanitary landfills	2.82	0	0
55a	Tons of residential refuse burned in single chamber incinerators	.00	200.	2.
55b	As 55a, except disposal in sanitary landfills	2.82	0	0
56a	Tons of waste burned on-site	.00	85.	280.
56b	As 56a, except compacted for strip mine landfill	11.16	0	0
57a	Tons of refuse burned at open dumps	.00	85.	280.
57b	As 57a, except converted to sanitary landfills	.93	0	0
58a	Tons of refuse subject to inter-mittent burning at landfill sites	.00	4.25	14.
58b	As 58a, except no burning	.72	0	0
59a	Tons of leaves burned on-site, which could not be mulched	.00	60.	280.
59b	As 59a, except collected by local government for landfill	4.10	0	0
60a	Tons of leaves burned on-site	.00	60.	280.
60b	As 60a, except mulched on-site	−40.00	0	0
61a	Tons of dried sewage sludge burned	.23	1.	.3
62a	Tons of green coffee beans processed	.06	0	0

(6) Nitrogen oxides (pounds) e_j^t	(7) Sulfur dioxide (pounds) e_j^t	(8) Particulate matter (pounds) e_j^t	(9) Benzo(a)-pyrene (thousand micrograms) e_j^t	(10) Source magnitude s_i	(11) Optimal activity level x_j^*
		10.2	6.2	357,000	0
		4.	520.	50,000	50,000
		.6	520.	50,000	0
		28.	0	33,600	33,600
		0	0	33,600	0
		15.	0	16,500	16,500
		0	0	16,500	0
		47.	232.	519,000	400,409
		0	0	519,000	118,591
		47.	232.	455,000	0
		0	0	455,000	455,000
.03	.06	2.35	11.6	706,000	706,000
0	0	0	0	706,000	0
1.	.8	47.	365.	30,600	0
0	0	0	0	30,600	30,600
1.	.8	47.	365.	3,400	0
0	0	0	0	3,400	3,400
2.	2.	1.4	0	97,000	97,000
0	0	2.6	0	36,000	36,000

Table 2.1 (Continued)

(1) Variable number j	(2) Activity unit and description	(3) Cost of abatement (dollars) C_j	(4) Carbon monoxide (pounds) e_j^4	(5) Hydro-carbons (pounds) e_j^5
62b	As 62a, with cyclone collectors	.38	0	0
63a	Tons of steel castings produced in uncontrolled electric arc furnaces	0	0	0
63b	As 63a, with baghouse collectors	2.90	0	0
64a	Tons of steel castings produced in open hearth furnaces	0	0	0
64b	As 64a, with venturi scrubbers	1.55	0	0
65a	Tons of steel castings produced in induction and controlled electric arc furnaces	1.45	0	0
66a	Tons of raw metal processed in gray iron foundries	0	0	0
66b	As 66a, with wet caps	1.10	0	0
66c	As 66a, with wet scrubber	3.30	0	0
66d	As 66a, with fabric filter	7.20	0	0
67a	Tons of metal processed in nonferrous foundries	0	0	0
68a	Tons of primary steel produced in electric arc furnaces	0	0	0
68b	As 68a, with fabric filters	.31	0	0
69a	Tons of primary steel produced in controlled basic oxygen furnaces	.14	0	0
69b	As 69a, with venturi scrubber	.28	0	0
70a	Tons of sinter produced with cyclone controls	.02	0	0
70b	As 70a, with electrostatic precipitators	.25	0	0
71a	Tons of iron produced in blast furnaces with primary dry cleaners	.07	0	0
71b	As 71a, with wet scrubbers	.16	0	0

(6) Nitrogen oxides (pounds) e_j^t	(7) Sulfur dioxide (pounds) e_j^t	(8) Particulate matter (pounds) e_j^t	(9) Benzo(a)-pyrene (thousand micrograms) e_j^t	(10) Source magnitude s_i	(11) Optimal activity level x_j^*
0	0	.1	0	36,000	0
0	0	10.6	0	20,300	20,300
0	0	.2	0	20,300	0
0	0	9.3	0	142,100	142,100
0	0	.8	0	142,100	0
0	0	.15	0	40,600	40,600
0	2.0	16.5	0	94,000	94,000
0	2.	4.1	0	94,000	0
0	2.	.8	0	94,000	0
0	2.	.3	0	94,000	0
0	0	5.2	0	25,600	25,600
0	0	10.6	0	1,000,000	0
0	0	.16	0	1,000,000	1,000,000
0	0	.40	0	2,250,000	2,250,000
0	0	.16	0	2,250,000	0
0	0	3.5	0	850,000	0
0	0	.2	0	850,000	850,000
0	0	5.4	0	1,417,500	0
0	0	.1	0	1,417,500	1,417,500

Data for the Linear Programming Model

Table 2.1 (Continued)

(1)	(2)	(3)	(4)	(5)
Variable number j	Activity unit and description	Cost of abatement (dollars) C_j	Carbon monoxide (pounds) e_j^t	Hydro-carbons (pounds) e_j^t
72a	Tons of grain processed through elevators	.04		0
72b	As 72a, with filter and aeration systems	.49		0
73a	Thousand barrels of petroleum feed through fluid catalytic cracking units			220.
73b	As 73a, with carbon monoxide waste heat boiler			11.
73c	As 73a, with electrostatic precipitator for catalytic dust			220.
73d	Combination of 73b and 73c			11.
73e	As 73a, with hydrotreater to desulfurize the fresh feed	ʃ		220.
73f	Combination of 73b and 73e	2		11.
73g	Combination of 73c and 73e	ʃ		220.
73h	Combination of 73d and 73e	ʃ		11.
74a	Thousand barrels of crude oil processed in refineries			90.
75a	Tons of dry fertilizer mixed and granulated			0
75b	As 75a, with wet scrubber			0
76a	Tons of nitric acid produced			0
76b	As 76a, with catalytic combuster			0
77a	Tons of ammonium nitrate produced			0
77b	As 77a, with wet scrubbers	.11		0
78a	Tons of sulfuric acid produced in plants built prior to 1967	.16		0
78b	As 78a, with double contact process	.98		0
79a	Tons of sulfuric acid produced in plants built after 1967	.16		0
79b	As 79a, with double contact process	.39		0

(6)	(7)	(8)	(9)	(10)	(11)
Nitrogen oxides (pounds) e'_j	Sulfur dioxide (pounds) e'_j	Particulate matter (pounds) e'_j	Benzo(a)-pyrene (thousand micrograms) e'_j	Source magnitude s_i	Optimal activity level x^*_j
0	0	6.	0	2,400,000	2,400,000
	0	.45		2,400,000	
	555.	60.		54,604	
	555.	60.		54,604	
	555.	.6		54,604	
	555.	.6		54,604	
	111.	60.		54,604	
	111.	60.		54,604	
	111.	.6		54,604	
	111.	.6		54,604	
	0	0		137,606	
	0	15.		200,000	
	0	.3		200,000	
	0	0		100,000	
2.75	0	0		100,000	
0	0	8.		125,000	
0	0	.4		125,000	125,000
0	45.	.4		1,110,000	0
0	4.5	.4		1,110,000	1,110,000
0	45.	.4		410,000	0
0	4.5	.4		410,000	410,000

Table 2.1 (Continued)

(1)	(2)	(3)	(4)	(5)
Variable number j	Activity unit and description	Cost of abatement (dollars) C_j	Carbon monoxide (pounds) e_j^*	Hydro-carbons (pounds) e_j^*
80a	Barrels of cement produced in plant with medium control efficiency	.10		0
80b	As 80a, with 99.6 percent efficient electrostatic precipitator	.21		0
80c	As 80a, with 99.9 percent efficient electrostatic precipitator	.27		0
81a	Barrels of cement produced in plant with low control efficiency	.07		0
81b	As 81a, with 98.4 percent efficient electrostatic precipitator	.20		0
81c	As 81a, with 99.5 percent efficient electrostatic precipitator	.23		0
82a	Barrels of cement produced in plants with high control efficiency	.14		0
82b	As 82a, with 99.8 percent efficient electrostatic precipitators	.17		0
83a	Tons of superphosphates produced	0·		0
84a	Tons of rock processed in asphaltic concrete plants	.02		0
85a	Tons of coal charged in coke ovens of steel mill	0		0
85b	As 85a, with wet scrubbers on charging cars	.23		0
86a	Tons of coal charged to produce foundry coke	0		0
86b	As 86a, with lowering sleeves and inspirating system	.01		0
86c	As 86a, with spray trucks to wet down coke stored in yards	.04		0
86d	Combination of 86b and 86c	.05		0
86e	As 86d, with dust collectors at transfer points	.15		0
86f	As 86e, with wet scrubbers on charging cars	.39		0

(6)	(7)	(8)	(9)	(10)	(11)
Nitrogen oxides (pounds) e_j^t	Sulfur dioxide (pounds) e_j^t	Particulate matter (pounds) e_j^t	Benzo(a)-pyrene (thousand micrograms) e_j^t	Source magnitude s_i	Optimal activity level x_j^*
0	0	1.008	0	3,760,000	3,760,000
0	0	.112	0	3,760,000	0
0	0	.039	0	3,760,000	0
0	0	1.96	0	2,455,000	2,455,000
0	0	.45	0	2,455,000	0
0	0	.15	0	2,455,000	0
0	0	.27	0	9,285,000	9,285,000
0	0	.09	0	9,285,000	0
0	0	17.5	0	6,250	6,250
0	0	.45	0	961,000	961,000
0	0	2.	0	1,326,000	1,326,000
0	0	.6	0	1,326,000	0
0	0	26.	0	440,000	0
0	0	25.4	0	440,000	0
0	0	19.	0	440,000	0
0	0	18.4	0	440,000	0
0	0	16.8	0	440,000	440,000
0	0	15.4	0	440,000	0

Data for the Linear Programming Model

Table 2.1 (Continued)

(1)	(2)	(3)	(4)	(5)
Variable number j	Activity unit and description	Cost of abatement (dollars) C_j	Carbon monoxide (pounds) e_j^t	Hydro-carbons (pounds) e_j^t
87a	Million cubic feet of coke oven by-product gas vented to atmosphere	0	0	10,615.
87b	As 87a, except sold as fuel to replace coal	−90.70	0	0
88a	Million cubic feet of stack gas flow from lead smelting	0	0	0
88b	As 88a, with recovery system to produce sulfuric acid	−1.14	0	0
89a	Million cubic feet of gas discharge from hydrocarbon processing	.86	0	87.
89b	As 89a, with afterburners	4.20	0	8.7
90a	Thousand tons of rock crushed and processed with primary controls	3.00	0	0
90b	As 90a, with cyclone collectors	6.00	0	0
91a	People using products which pollute	0	0	2.35
92a	Tons of industrial solvents used	0	0	2,000.
92b	As 92a, with afterburners	108.00	0	200.
93a	Tons of dry cleaning solvents used	0	0	2,000.
94a	Tons of nonindustrial solvents used	0	0	2,000.
95a	Thousand gallons of gasoline stored in floating roof tanks	.14	0	100.8
96a	Thousand gallons of gasoline stored in cone roof tanks built before 1969	0	0	465.8
96b	As 96a, except converted to floating roof tanks	.23	0	100.8
97a	Thousand gallons of gasoline stored in cone roof tanks built after 1968	0	0	465.8
97b	As 97a, except with floating roofs built in as original equipment	.14	0	100.8

(6) Nitrogen oxides (pounds) e_j^t	(7) Sulfur dioxide (pounds) e_j^t	(8) Particulate matter (pounds) e_j^t	(9) Benzo(a)-pyrene (thousand micrograms) e_j^t	(10) Source magnitude s_i	(11) Optimal activity level x_j^*
0	0	0	0	1,450	0
0	0	0	0	1,450	1,450
0	800.	2.	0	218,000	0
0	160.	0	0	218,000	218,000
0	0	0	0	432,000	432,000
0	0	0	0	432,000	0
0	0	5,000.	0	4,000	0
0	0	2,750.	0	4,000	4,000
0	0	3.65	0	2,600,000	2,600,000
0	0	0	0	24,800	24,800
0	0	0	0	24,800	0
0	0	0	0	5,000	5,000
0	0	0	0	4,820	4,820
0	0	0	0	868,000	868,000
0	0	0	0	226,000	0
0	0	0	0	226,000	226,000
0	0	0	0	64,000	0
0	0	0	0	64,000	64,000

Data for the Linear Programming Model

measured in the units identified in column 2. These magnitudes are extrapolations of 1968 levels, determined from census data and discussions with industry representatives. The abatement activities for each source are mutually exclusive; they are so defined and combined that only the *sum* of activity levels can equal the corresponding source magnitude. The successive elements of the **s** vector are $s_1 = 330,700$; $s_2 = 180,800$; $s_3 = 625,500$; $s_4 = 7,252$; $s_5 = 69,350$; etc. The pollution sources include the various modes of transportation (source numbers 1 through 10), the combustion of coal, oil, and gas in stationary furnaces and boilers (numbers 11 through 51), the burning of refuse (numbers 52 through 61), industrial processes (numbers 62 through 90), and evaporation and miscellaneous sources (numbers 91 through 97). This categorization of pollution sources is for the most part based on a study of the St. Louis airshed prepared by Venezia and Ozolins (1967).

Column 3 of table 2.1 contains the unit cost of abatement for each of the 250 activities. In those cases in which the initial activity represents nonabatement, the unit cost is zero. In the preceding chapter, unit costs were defined in terms of a single input, labor. Here costs are the dollar value of labor, materials, and capital. The calculation of a unit cost of an abatement activity is illustrated with 73b. This particular activity represents the operation of a carbon monoxide waste heat boiler in petroleum refineries equipped with fluid catalytic cracking units.

The estimated cost for installing a carbon monoxide waste heat boiler on such a unit with a petroleum feed capacity of 40,000 barrels a day is $1,770,000.[3] The estimated annual costs (in dollars) are as follows:

Depreciation at 5 percent	88,500	
Opportunity cost of capital, 10 percent	177,000	
Maintenance and other costs	83,000	
Total annual costs	348,500	(2.10)
Less value of recovered steam	315,000	
Net annual cost	33,500	

Because the latter is equivalent to 14,600 thousand barrels a year, the cost per source unit is

$$\$33,500/14,600 \cong \$2.30 \text{ per thousand barrels of petroleum feed.} \quad (2.11)$$

Some simplifying assumptions are inherent in these calculations. The

66

opportunity cost of capital represents the value of output foregone as a consequence of the investment in pollution control equipment. A uniform rate of 10 percent is used on the assumption that the net return on new investment, corrected for relative risk, is the same for all industries. (A different approach, based on Baumol's, 1967, interpretation of the law of large numbers, would have been to use a higher opportunity cost of capital rate for more profitable industries.)

The costs in (2.10) are assumed to be uniform over the life of the equipment. In fact, as the machinery is depreciated, capital becomes available for new investment, allowing for reductions in the opportunity cost of the initial investment. To avoid this complication it could be assumed that the depreciation is exactly offset by increments of new investment in the same equipment.

The implication in (2.11) is that the investment is perfectly divisible. In fact, it is likely that unit costs vary with the capacity of the installation.

The costs in (2.10) represent the value of output that these inputs would produce in their next best alternative use. These are social or economic costs rather than private or accounting costs. An example of a deviation in social and private costs is contained in the data in table 2.1. The market costs of substituting natural gas for coal (activities 17e, 18e, 19e, etc.) are augmented by a surcharge that reflects the relative scarcity of natural gas. This surcharge is the estimated increase in price that consumers, at some time in the future, would have to pay for synthetic gas made from coal, to replace natural gas used in 1975 for air pollution control, that is, 18¢ per thousand cubic feet of natural gas.

Some of the costs in table 2.1 are negative, or simply less than the initial cost. This indicates that there were opportunities for increasing private profits, which had gone unnoticed. For example, there is the case (see activities 25b and 87b) in which a coke producer could pipe waste by-product gas, otherwise vented to the atmosphere as pollution, to a neighboring manufacturer for whom it would provide a less polluting fuel than coal.

For each process, total cost and total emissions are assumed to be proportional to the activity level of that process. However, the model does allow for increasing marginal costs of abatement, as represented by sequences of activities (such as 86a, 86c, 86d, 86e, and 86f) in which unit costs increase more rapidly than some pollutant emissions decrease.[4]

Columns 4 through 9 of table 2.1 contain elements of the (6 × 250)

Table 2.2

Projected emission flows in the St. Louis airshed in 1975 in the absence of air pollution regulations, allowable annual flows, and abatement requirements (millions of pounds)

Category of source	Carbon monoxide	Hydro-carbons	Nitrogen oxides	Sulfur dioxide	Partic-ulates	Benzo(a)-pyrene[a]
Transportation	3,327	717	149	14	26	338
Combustion of fuel oil	(less than 0.5)	1	33	55	4	8
Combustion of coal by residential, industrial, and commercial users	9	2	31	191	48	90
Combustion of coal by residential users	23	5	4	54	22	427
Combustion of coal by public utilities	3	1	132	755	34	13
Combustion of natural and by-product gases	(less than 0.5)	125	56	45	6	22
Combustion of refuse	92	292	2	2	57	275
Industrial processes	748	77	9	273	94	0
Evaporation and miscellaneous minor sources	0	298	0	0	9	0
Total flows in the absence of regulations	**4,202**	**1,518**	**416**	**1,389**	**300**	**1,173**
Allowable flows	2,335	994	304	400	136	633
Required abatement	**1,867**	**524**	**112**	**989**	**164**	**540**

Note: The emission flows were projected from data collected in 1968.

[a]Benzo(a)pyrene emissions are in thousands of grams.

matrix **e**. These entries denote pounds of carbon monoxide, hydrocarbons, nitrogen oxides, sulfur dioxide, particulate matter, and thousands of micrograms (milligrams) of benzo(a)pyrene emitted per activity unit of each process. In general, emissions of each pollutant from a particular source are less for more costly processes. However, the reverse may be true; the control devices (see activities 1b and 2b) to reduce carbon monoxide and hydrocarbon emissions for pre-1970 automobiles were expected to *increase* nitrogen oxide emissions. In many cases, for more costly processes, emissions of a single pollutant decrease while emissions of the remaining pollutants are the same.

The table does not indicate the elements, all 1's and 0's, of the (97 × 250) matrix **u**. The reader may, for example, confirm that the elements $u_{5,5a}$ and $u_{5,5b}$, which relate activities 5a and 5b to pollution source 5, are both 1's. The remaining 248 elements, $u_{5,j}$, are all 0's.

All that remains to complete the input for the Linear Programming Model is the set of allowable annual emission flows. These levels, which are the elements of the (6 × 1) vector **f**, are as follows:

f^c = 2,335,162,803 pounds of carbon monoxide;
f^h = 994,455,118 pounds of hydrocarbons;
f^n = 303,532,644 pounds of nitrogen oxides;   (2.12)
f^s = 400,424,669 pounds of sulfur dioxide;
f^p = 135,802,509 pounds of particulates;
f^b = 633,055,870 thousand micrograms of benzo(a)pyrene.

These allowable annual flows are based upon a linear relationship between total emission flows and pollutant concentrations, which will be explained in chapter 4. It should be noted that the model is simplified by aggregating all hydrocarbons, regardless of chemical composition, and all particulates, whether submicron is size or greater, into single categories. In theory, each pollutant category should represent homogeneous units.

The solution of the Linear Programming Model includes an optimal vector of activity levels, $\mathbf{x}^*$. To make table 2.1 more meaningful, the individual elements x_j^* of that vector are included in column 11 of the table. This solution will be discussed at length in chapter 3.

Emissions in the Absence of Regulations

The data in table 2.1 can be used to determine total pollution flows for the year 1975 in the absence of a regulatory program. Letting $\mathbf{x}_a$ denote the vector of annual activity levels in which $x_{1a} = s_1$, $x_{2a} = s_2$, $x_{3a} = s_3$, ..., $x_{97a} = s_{97}$, and in which all of the alternative abatement activity levels, the x_{jb}, x_{jc}, etc., are equal to zero, the projected emission flows $\mathbf{f}_a$, are

$$\mathbf{f}_a = \mathbf{e}\mathbf{x}_a. \tag{2.13}$$

The total flows, subdivided into pollution source categories, are shown in table 2.2. It should be emphasized that this projection of annual emission flows assumes a continuation of the base-year level of abatement, which in this case is the year 1963. This represents abatement undertaken because of nuisance ordinances, for purposes of community good will, for the benefit of employees, or for other private cost considerations.

The base-year level of abatement is not insignificant. For example, the projected particulate emissions from Granite City Steel Company (source numbers 69, 70, 71, 85, and portions of 11 and 51) in 1975 total 14 million pounds. Without any of the base-year control methods, the particulate emissions from this company alone would have exceeded 400 million pounds.[5]

When the allowable annual emissions flows are deducted from the projected flows, the required abatement quantities are determined.[6] These appear at the bottom of table 2.2.

Quality of the Data

The data from table 2.1 and the requirement set (2.12) are used to implement Model I. Later chapters of the book include alternative versions of the model incorporating different or additional data.

For the reader who is planning to implement in his own airshed a model similar to one of those listed in table 1.1, several comments may be added. Given the uncertainties in the data, it would have been preferable to round out the emission factors and allowable flows. As can be seen in table 2.1, some of the emission factors are carried to three significant figures to the right of the decimal point, while in the requirement set (2.12) the allowable flows are carried to the nearest pound. Had there been more rounding out of these numbers, the results would have changed relatively little.

The prospective model builder will have advantages in assembling data. The parameters here were calculated in 1968, when there was relatively little work in print on abatement costs. As a consequence, many of the numbers had to be estimated in the course of private conversations with industry and control agency representatives. Since that time more reliable data have been published in government documents and scientific journals. In addition, regulatory agencies are now required to maintain emission inventories. This was not the case when the data in table 2.1 were collected, and as a result certain sources were not known to the writer.[7] It should be emphasized that table 2.1 is included here to illustrate the organization of material for the Linear Programming Model; it should not be used as a source for current cost and engineering data.

The current availability of good emission inventories permits the model builder to better distinguish significant categories of pollution sources. For example, when Model I was implemented for the St. Louis airshed, there were no data for segregating some very important hydrocarbon processing industries, such as manufacturers of soap, organic and inorganic chemicals, beer, etc. Consequently, they were aggregated under a single source, 89, in this model. On the other hand, the results of this model suggest that there can be too much disaggregation, as in the excessive delineation of types of industrial and commercial coal burning furnaces.

It should be emphasized that the Linear Programming Model must be distinguished from the specific data for the St. Louis airshed with which it was implemented. The more reliable the data and the more skillfully they are organized, the more useful the results will be. What is demonstrated in this book is the substantial power of the model for dealing with complex policy issues.

APPENDIX: NUMERICAL EXAMPLE

The following simple example illustrates Model I. Consider a hypothetical airshed with a single polluting industry, cement manufacturing.[8] Annual production is 2,500,000 barrels of cement. Currently, the five kilns, each of which produces 500,000 barrels of output, are equipped with mechanical collectors for air pollution control and are emitting 2 pounds of particulate matter for every barrel of cement produced. The emission rate can be reduced from 2.0 to .5 pounds of patriculates by replacing the mechanical collectors with 97.5 percent efficient electrostatic precipitators, or to .2 pounds of particulates per barrel of cement by replacing the mechanical collectors with 99 percent efficient electrostatic precipitators. The unit costs of abatement for the mechanical collector and the two alternative electrostatic precipitators, per barrel of cement produced, are $.06, $.20, and $.24, respectively. The current annual flow of particulates in this airshed, 5 million pounds per year, is considered excessive by the regulatory agency, which has determined that the allowable flow should be 800,000 pounds per year. How can this goal be achieved at the least total cost?

This problem is represented by the following linear program:

$$\text{Minimize } Z = \$.06x_{1a} + \$.20x_{1b} + \$.24x_{1c}$$

$$\text{subject to} \quad
\begin{aligned}
x_{1a} + \quad & x_{1b} + \quad x_{1c} = 2,500,000, \\
2.0x_{1a} + \quad & .5x_{1b} + \quad .2x_{1c} \leq 800,000, \\
x_{1a}, \quad & x_{1b}, \quad x_{1c} \geq 0.
\end{aligned}
\qquad (2.14)$$

The least-cost combination of abatement activities and the minimized total cost of abatement are as follows:[9]

$$
\begin{aligned}
x_{1b} &= 1,000,000 \text{ barrels of cement,} \\
x_{1c} &= 1,500,000 \text{ barrels of cement,} \\
Z &= \$560,000 \text{ a year.}
\end{aligned}
\qquad (2.15)$$

The objective could be achieved by requiring 97.5 percent efficient electrostatic precipitators on two of the kilns and 99 percent efficient electrostatic precipitators on the remaining three kilns, or by establishing a maximun emission rate of .32 pounds of particulates per barrel of cement produced. Using determinantal ratios, the reader may confirm that the shadow prices are $\$(-2/15)$ per pound of particulates and $\$(31/150)$ per barrel of cement. The former is the reduction in the total cost of abatement if the allowable level of particulates were increased to 800,001. The latter is the increase in the total cost of abatement if the production magnitude rose to 2,500,001.

3

SOLUTION OF
THE EMPIRICAL MODEL

The solution of the Linear Programming Model includes a total annual cost of abatement, shadow prices, and a set of efficient abatement activity levels.[1] Some details of the latter are presented in this chapter, particularly as they apply to industrial coal furnaces. Because many of the numbers in the model are estimates, sensitivity analysis becomes an important tool; and several of its applications are illustrated in this chapter. The optimal solution is contrasted with an equiproportional solution and to the legal solution in the St. Louis airshed. Next, the model is used to examine an allegation that a least-cost set of air pollution control activities is likely to significantly augment the flows of joint-wastes such as water pollution, thermal pollution, and solid waste. The usefulness of the model as a guide for regulatory agencies is considered, and finally an effort is made to verify the estimate of the total cost of abatement.

We shall, from time to time, relate the present model to other empirical models of air quality management and draw comparisons that may be useful.

Total Cost of Abatement

The total cost of abatement, given the input data in table 2.1 and the allowable emission flows of the six pollutants, is the scalar product Cx_I^*, which equals \$46,525,242. (The asterisk denotes the optimal set of activity levels, and the subscript indicates Model I.) The breakdown of this total cost by pollution source categories is contained in table 3.1. Assuming that there were no abatement beyond the initial or base-year effort, the total cost of abatement would then be Cx_a, where x_a is the vector defined in chapter 2, in which the alternative abatement activity variables are equal to zero. This total cost would be \$11,187,959 in 1975.[2] The incremental cost of abatement for reducing emissions to the given allowable flows is therefore $Cx_I^*-Cx_a$, \$35,337,283 per year.[3] It is of interest to note here that this book contains four tables, each giving a different kind of breakdown of this same total incremental cost of abatement.

Other investigators have developed empirical models which, like Model I, are based on reducing total annual emissions to a predetermined set of allowable flows. These models are listed in table 3.2.

Table 3.1

Annual cost of controlling air pollution in the St. Louis airshed in 1975

Category	Source numbers	Base-year costs (dollars)	Incremental costs (dollars)	Total costs (dollars)
Transportation	1–10	0	16,536,917	16,536,917
Combustion of fuel oil	11–16	0	99,250	99,250
Combustion of coal by industrial and commercial users	17–31 34–37	811,806	3,762,888	4,574,694
Combustion of coal by residential users	32,33	0	1,388,720	1,388,720
Combustion of coal by public utilities	38–44	5,156,280	10,115,603	15,271,883
Combustion of natural and by-product gases	45–51	1,753,792	0	1,753,792
Combustion of refuse	52–61	22,310	1,736,086	1,758,396
Industrial processes	62–90	3,322,251	1,636,879	4,959,130
Evaporation and miscellaneous minor sources	91–97	121,520	60,940	182,460
Totals		11,187,959	35,337,283	46,525,242

Note: This breakdown of abatement costs is derived from the solution of Model I.

Shadow Prices

It was demonstrated in chapter 1 that shadow prices have theoretical significance for environmental decision making. A comparison of a pollutant shadow price with the marginal damage caused by that pollutant can indicate whether the corresponding air quality standard is sufficiently stringent. Alternatively, if only the relative harmfulness of each individual pollutant is known, the *ratio* of pollutant shadow prices may be significant. Furthermore, we shall see in chapter 7 that the pollutant shadow prices have a useful interpretation as Pigouvian fees.

The solution of the Linear Programming Model includes a set of dual values or shadow prices.[4] The shadow price associated with a specific source magnitude s_i or emission flow f^j is the *change* in total cost $\mathbf{Cx}^*$ associated with a unit increase in that constraint. Table 3.3 contains the shadow prices corresponding to the optimal solution of Model I. Note that the total emission flow of benzo(a)pyrene in the optimal solution is 572,068,300 milligrams, which is less than the allowable flow. Accordingly, the benzo(a)pyrene constraint is not binding and the shadow price associated with this pollutant is zero. In subsequent versions of the Linear Programming Model, the benzo(a)pyrene requirement was deleted. This was done, not only because the requirement was not binding but also because this pollutant was no longer being monitored in the St. Louis airshed.

Although the shadow prices are generated by the computer, it is useful to observe how they may be calculated independently. The determinantal ratio (3.1), on p.78, is equivalent to the shadow price for carbon monoxide. The denominator of this ratio is an abbreviation of the (102×102) basis matrix $\mathbf{B}$ whose first row is the sequence of carbon monoxide emission factors for the *nonzero* activities, including the zero activity level for source 13. There are 102 such activities $(x_{1a}, x_{2a}, x_{3b}, x_{3c}, x_{3d}, x_{4a}, x_{5a}, \cdots, x_{95a}, x_{96b}, x_{97b})$, the same as the number of binding constraints (97 pollution sources plus 5 emission flows). The second row is the sequence of hydrocarbon emission factors. Because the benzo(a)pyrene requirement is not binding, none of the emission factors for this pollutant are included in the $\mathbf{B}$ matrix. The sixth row contains those elements $(u_{1,1a}, u_{1,2a}, u_{1,3b}, u_{1,3c}, u_{1,3d}, u_{1,4a}, \cdots, u_{1,97b})$ from the first row of the $\mathbf{u}$ matrix for which the corresponding activity is nonzero. Accordingly, the final 97 rows of the $\mathbf{B}$ matrix are the (97×102) subset of the (97×250) matrix $\mathbf{u}$. The numerator of (3.1) is exactly like the denominator

Table 3.2

Empirical models of air quality management based on total allowable emission flows

Model	Jackson and Wohlers
Empirical application	Delaware Valley region (1980 and 2000)
Air quality indicator	Annual emissions of sulfur dioxide, particulates, carbon monoxide, and hydrocarbons
Objective	Minimize total cost of pollution abatement
Types of sources	Mobile, point, and area sources
Abatement alternatives	Control methods for all types of sources
Reference	W. E. Jackson and H. C. Wohlers, "Regional Air Pollution Control Costs," *Journal of the Air Pollution Control Association*, 22, September, 1972, pp. 679–684.

Siegel, Ehrenfeld, and Morganstern	Atkinson and Lewis (ELC)
Boston area (1973)	St. Louis region
Annual emissions of particulates and sulfur dioxide	Annual emissions of particulates
Select a control strategy on the basis of percentage emission reductions, technical feasibility, economic impact, and enforcement feasibility	Minimize total cost of pollution abatement
Mobile, point, and area sources	27 largest point sources of particulate emissions accounting for approximately 80 percent of total particulate emissions
Ten strategies (sets of control methods, such as fuel substitutions, fuel washing, annual inspections, etc.) for fuel burning sources only	Scrubbers, dry collectors, electrostatic precipitators, fuel substitutions, etc.
Richard D. Siegel, John Ehrenfeld, and Paul Morganstern, "A Strategy for Reduction of Particulate Emissions in the Boston Area," *Journal of the Air Pollution Control Association, 25,* March, 1975, pp. 256–259.	Scott E. Atkinson and Donald H. Lewis, "A Cost-Effectiveness Analysis of Alternative Air Quality Control Strategies," *Journal of Environmental Economics and Management, 1,* November, 1974, pp. 237–250.

$$\pi^C = \begin{array}{|ccccccccc|}
0 & 0 & 10.25 & 28.71 & 30.76 & . & . & . & .14 \\
616 & 616 & 616 & 212.52 & 129.72 & . & . & . & 100.8 \\
113 & 113 & 22.6 & 22.6 & 22.6 & . & . & . & 0 \\
9 & 9 & 9 & 9 & 9 & . & . & . & 0 \\
11 & 11 & 11 & 11 & 11 & . & . & . & 0 \\
1 & 0 & 0 & 0 & 0 & . & . & . & 0 \\
0 & 1 & 0 & 0 & 0 & . & . & . & 0 \\
0 & 0 & 1 & 1 & 1 & . & . & . & 0 \\
0 & 0 & 0 & 0 & 0 & . & . & . & 0 \\
. & . & . & . & . & . & . & . & . \\
. & . & . & . & . & . & . & . & . \\
. & . & . & . & . & . & . & . & . \\
0 & 0 & 0 & 0 & 0 & . & . & . & 1 \\ \hline
2910 & 2910 & 2910 & 931.2 & 931.2 & . & . & . & 0 \\
616 & 616 & 616 & 212.52 & 129.72 & . & . & . & 100.8 \\
113 & 113 & 22.6 & 22.6 & 22.6 & . & . & . & 0 \\
9 & 9 & 9 & 9 & 9 & . & . & . & 0 \\
11 & 11 & 11 & 11 & 11 & . & . & . & 0 \\
1 & 0 & 0 & 0 & 0 & . & . & . & 0 \\
0 & 1 & 0 & 0 & 0 & . & . & . & 0 \\
0 & 0 & 1 & 1 & 1 & . & . & . & 0 \\
0 & 0 & 0 & 0 & 0 & . & . & . & 0 \\
. & . & . & . & . & . & . & . & 0 \\
. & . & . & . & . & . & . & . & 0 \\
. & . & . & . & . & . & . & . & 0 \\
0 & 0 & 0 & 0 & 0 & . & . & . & 1
\end{array} \qquad (3.1)$$

except that the first row, containing carbon monoxide emission factors, is replaced by the vector of costs associated with the nonzero activities. Thus, this ratio determines the shadow price for the pollutant carbon monoxide. These determinants may be successively reduced to determinants of smaller order until a single fraction is obtained, as in equation (3.2).

While it is obvious that the unit cost and emission factors determine what activities will be nonzero, it should also be clear that the pollutant shadow prices are based exclusively on costs and emission factors associated with intermediate activity levels (that is, activities ij for which $0 < x_{ij} < s_i$).

$$
\frac{\begin{vmatrix}
10.25 & 28.71 & 30.76 & 1.98 & 8.34 & .83 & 2.08 & 0 & 11.16 \\
616 & 212.52 & 129.72 & 1 & 9.216 & 1 & 1 & 280 & 0 \\
22.6 & 22.6 & 22.6 & 20 & 4.518 & 20 & 20 & .6 & 0 \\
9 & 9 & 9 & 60.8 & .008 & 117.8 & 60.8 & 1.2 & 0 \\
11 & 11 & 11 & 2.5 & .38 & 2.5 & 2.5 & 47 & 0 \\
1 & 1 & 1 & 0 & 0 & 0 & 0 & 0 & 0 \\
0 & 0 & 0 & 1 & 1 & 0 & 0 & 0 & 0 \\
0 & 0 & 0 & 0 & 0 & 1 & 1 & 0 & 0 \\
0 & 0 & 0 & 0 & 0 & 0 & 0 & 1 & 1
\end{vmatrix}}{\begin{vmatrix}
2910 & 931.2 & 931.2 & 3 & 3 & 3 & 3 & 85 & 0 \\
616 & 212.52 & 129.72 & 1 & 9.216 & 1 & 1 & 280 & 0 \\
22.6 & 22.6 & 22.6 & 20 & 4.518 & 20 & 20 & .6 & 0 \\
9 & 9 & 9 & 60.8 & .008 & 117.8 & 60.8 & 1.2 & 0 \\
11 & 11 & 11 & 2.5 & .38 & 2.5 & 2.5 & 47 & 0 \\
1 & 1 & 1 & 0 & 0 & 0 & 0 & 0 & 0 \\
0 & 0 & 0 & 1 & 1 & 0 & 0 & 0 & 0 \\
0 & 0 & 0 & 0 & 0 & 1 & 1 & 0 & 0 \\
0 & 0 & 0 & 0 & 0 & 0 & 0 & 1 & 1
\end{vmatrix}}
$$

$$
= \frac{\begin{vmatrix}
18.46 - 20.51 \begin{bmatrix} 403.48 \\ 486.28 \end{bmatrix} & 0 & 0 & 0 & 0 \\
0 & -486.28 & 0 & 0 & 0 \\
0 & 0 & -15.482 & 0 & .6 \\
0 & 0 & 0 & -57 & 0 \\
0 & 0 & -2.12 & 0 & 47
\end{vmatrix}}{\begin{vmatrix}
-1978.8 & 0 & 0 & 0 & 0 \\
0 & -82.8 & 0 & 0 & 0 \\
0 & 0 & -15.482 & 0 & .6 \\
0 & 0 & 0 & -57 & 0 \\
0 & 0 & -2.12 & 0 & 47
\end{vmatrix}}
$$

$$
= \frac{\left(18.46 - 20.51 \begin{bmatrix} 403.48 \\ 486.28 \end{bmatrix}\right)(-486.28)}{(-1978.8)(-82.8)} = -.00428. \tag{3.2}
$$

Solution of the Empirical Model

Table 3.3
Shadow prices in the solution of the Linear Programming Model

Source number*	Source magnitude	Shadow price (dollars)	Product (dollars)	Cumulative total of products (dollars)
1	330,700	65.63964	21,707,028	21,707,028
2	180,800	65.63964	11,867,646	33,574,674
3	625,500	46.38387	29,013,110	62,587,784
4	7,252	86.57237	627,823	63,215,607
5	69,350	86.57237	6,003,794	69,219,401
6	136,500	10.57825	1,443,931	70,663,332
7	14,700	.47373	6,964	70,670,296
8	2,100	1.89342	3,976	70,674,272
9	52,500	3.29575	173,027	70,847,299
10	4,200	8.01632	33,669	70,880,968
11	39,700	40.63955	1,613,390	72,494,358
12	158,800	39.74754	6,311,909	78,806,267
13	0	40.72622	0	78,806,267
14	11,540	25.40864	293,216	79,099,483
15	150,100	25.40864	3,813,837	82,913,320
16	8,017	25.40864	203,701	83,117,021
17	580	7.55484	4,382	83,121,403
18	2,400	8.09484	19,428	83,140,831
19	173,000	10.07246	1,742,536	84,883,367
20	358,800	10.17246	3,649,879	88,533,246
21	97,600	10.15246	990,880	89,524,126
22	114,000	10.58670	1,206,884	90,731,010
23	444,000	10.79670	4,793,735	95,524,745
24	110,000	9.69439	1,066,383	96,591,128
25	29,000	3.45388	100,163	96,691,291
26	35,000	8.86733	310,357	97,001,648
27	63,500	8.71733	553,550	97,555,198
28	8,000	8.09992	64,799	97,619,997
29	17,000	8.28992	140,929	97,760,926
30	28,000	8.36992	234,358	97,995,284
31	15,000	8.30992	124,649	98,119,933

Source number*	Source magnitude	Shadow price (dollars)	Product (dollars)	Cumulative total of products (dollars)
32	1,120	3.15724	3,536	98,123,469
33	428,000	3.91231	1,674,469	99,797,938
34	32,000	3.97220	127,110	99,925,048
35	24,000	8.05941	193,426	100,118,474
36	60,000	7.69348	461,609	100,580,083
37	44,000	7.01158	308,510	100,888,593
38	1,000,000	6.91769	6,917,690	107,806,283
39	200,000	10.16619	2,033,238	109,839,521
40	132,000	10.39803	1,372,540	111,212,061
41	730,000	7.24967	5,292,259	116,504,320
42	3,000,160	4.99295	14,979,648	131,483,968
43	3,430,000	3.87384	13,287,271	144,771,239
44	2,070,000	1.55000	3,208,500	147,979,739
45	10,500	139.25826	1,462,212	149,441,951
46	123,000	82.04756	10,091,849	159,533,800
47	112,500	50.13870	5,640,604	165,174,404
48	38,000	50.13870	1,905,271	167,079,675
49	7,262	98.09667	712,378	167,792,053
50	26,176	147.18000	3,852,584	171,644,637
51	17,974	59.47769	1,069,052	172,713,689
52	357,000	2.14995	767,532	173,481,221
53	50,000	1.01051	50,526	173,531,747
54	33,600	2.43688	81,879	173,613,626
55	16,500	2.40302	39,650	173,653,276
56	519,000	11.16000	5,792,040	179,445,316
57	455,000	.93000	423,150	179,868,466
58	706,000	.55800	393,948	180,262,414
59	30,600	4.10000	125,460	180,387,874
60	3,400	−40.00000	−136,000	180,251,874
61	97,000	1.04682	101,542	180,353,416
62	36,000	.26145	9,412	180,362,828

Table 3.3 (Continued)

Source number*	Source magnitude	Shadow price (dollars)	Product (dollars)	Cumulative total of products (dollars)
63	20,300	.82130	16,672	180,379,500
64	142,100	.72058	102,394	180,481,894
65	40,600	1.46162	59,342	180,541,236
66	94,000	1.32230	124,296	180,665,532
67	25,600	.40290	10,314	180,675,846
68	1,000,000	.32240	322,400	180,998,246
69	2,250,000	.17099	384,728	181,382,974
70	850,000	.26550	225,675	181,608,649
71	1,417,500	.16775	237,786	181,846,434
72	2,400,000	.50489	1,211,736	183,058,170
73	54,604	42.88715	2,341,810	185,399,980
74	137,606	3.93826	541,928	185,941,908
75	200,000	.58324	116,648	186,058,556
76	100,000	1.44758	144,758	186,203,314
77	125,000	.14099	17,624	186,220,938
78	1,110,000	1.10968	1,231,745	187,452,683
79	410,000	.51968	213,069	187,665,752
80	3,760,000	.17810	669,656	188,335,408
81	2,455,000	.22186	544,666	188,880,074
82	9,285,000	.16092	1,494,142	190,374,216
83	6,250	1.35593	8,475	190,382,691
84	961,000	.05487	52,730	190,435,421
85	1,326,000	.15496	205,477	190,640,898
86	440,000	1.45169	638,744	191,279,642
87	1,450	-90.70000	-131,515	191,148,127
88	218,000	2.36877	516,392	191,664,519
89	432,000	3.01399	1,302,044	192,966,563
90	4,000	219.07415	876,297	193,842,860
91	2,600,000	.34099	886,574	194,729,434
92	24,800	49.51691	1,228,019	195,957,453
93	5,000	49.51691	247,585	196,205,038

Source number*	Source magnitude	Shadow price (dollars)	Product (dollars)	Cumulative total of products (dollars)
94	4,820	49.51691	238,672	196,443,710
95	868,000	2.63565	2,287,744	198,731,454
96	226,000	2.72565	615,997	199,347,451
97	64,000	2.63565	168,682	199,516,133
CO	2,335,162,803	−.00428	−9,994,497	189,521,636
HC	994,455,118	−.02476	−24,622,709	164,898,927
NO	303,532,644	−.32639	−99,070,019	65,828,908
SO	400,424,669	−.02193	−8,781,313	57,047,595
PM	135,802,509	−.07748	−10,521,979	46,525,616
BP	572,068,300	−.00000	0	46,525,616

*The sources are defined in table 2.1.

Every other activity in the basis vector is characterized by a u_{ij} element, which is the only 1 in its particular row of the **B** matrix. Accordingly, in the calculation of shadow prices from determinantal ratios, these other activities are eliminated in the process of row and column reduction.

The shadow price of each pollutant for which the allowable level is binding may be derived with a determinantal ratio comparable to (3.1). From table 3.3 it can be seen that these five shadow prices, which represent the costs of eliminating one additional pound of the respective pollutants, differ from one another by as much as several orders of magnitude. However, there is less disparity in the cost of reducing each of the allowable flows by some common small percent. Assuming that the shadow prices are constant within the neighborhood of, say, ± .001 of the allowable flows, the cost of reducing the allowable flows by .1 percent would be:

$$(.00428) \ (2,335,163) \cong \$10,000 \text{ for carbon monoxide,}$$
$$(.02476) \ (994,455) \ \ \cong \$25,000 \text{ for hydrocarbons,}$$
$$(.32639) \ (303,533) \ \cong \$99,000 \text{ for nitrogen oxides,} \qquad (3.3)$$
$$(.02193) \ (400,425) \ \cong \$ \ 9,000 \text{ for sulfur dioxide,}$$
$$(.07748) \ (135,803) \ \cong \$11,000 \text{ for particulates.}$$

The shadow price for each of the 97 pollution sources represents the increased cost of pollution abatement if that source magnitude were increased by one unit. This includes C_j, the cost of abatement borne by that source, plus $(-\sum_i \frac{5}{i}\pi^i e^i_j)$, the additional costs of abatement that would be required of other sources in the airshed in order to eliminate the incremental emissions. Although the shadow prices for pollution sources may be calculated with determinantal ratios, it is easier to use a formula corresponding to (1.34).

One of the properties of a linear programming model is that the dot products of shadow prices and constraints sum to the value of the objective function. In this case,

$$\sum_{i=1}^{97} \pi_i s_i + \sum_{j=1}^{6} \pi^j q^j = C x^*_I = \$46,525,242. \tag{3.4}$$

This summation is demonstrated by the cumulative totals in table 3.3. The final sum is slightly higher than the correct sum because of rounding errors.

The Control Method Solution

The least-cost set of abatement activities for Model I is represented by the vector x^*_I. This vector contains 101 nonzero activity levels, of which 54 are alternative activities. The total costs associated with the latter are summarized in table 3.4. We shall examine certain details of the optimal solution for coal burning furnaces.

At the time that Model I was implemented, Peabody Coal Company was suing the Missouri Air Conservation Commission for banning high sulfur coal (see Kohn, 1971a). Among the charges was the allegation that the use of low sulfur coal was not economical. The results of Model I indicate that low sulfur coal should be burned in some stokers but not in others. Specifically, a large proportion of the coal used in chain grate stokers should have a reduced sulfur content (see 19d, 20d, and 21d in table 3.4).[5] This is not the case for pulverized coal units, which are generally equipped with electrostatic precipitators for collecting particulate matter. The electrostatic precipitators are less effective when low sulfur coal is burned, and consequently the decrease in sulfur dioxide emissions is accompanied by an increase in particulates (compare, for example, activities 23b and 23c in table 2.1). The potential inefficiency of low sulfur coal in combination with electrostatic precipitators is accounted for in the Linear Programming Model.

There is another problem involved in requiring low sulfur coal for pulverized units. This occurs if, to avoid the restriction on sulfur content, the operator of the furnace converts to natural gas. Furnaces burning pulverized coal have a relatively small handling cost and achieve a higher heating efficiency with coal than with natural gas. For such furnaces, the pollution control benefits of natural gas are less than the incremental costs of conversion. The fact that the low sulfur restrictions did stimulate some conversions from pulverized coal to natural gas is indicative of economic inefficiency.

Control officials in the St. Louis airshed anticipated, correctly, that the low sulfur restrictions would accelerate the shift from coal to natural gas. With the exception of the pulverized coal furnaces, their expectations in this regard were consistent with the optimal solution of Model I.

One drawback to the use of natural gas for air pollution control caused some concern. It was alleged that natural gas is a direct source of air pollution because of leaking valves and pipelines.[6] Although the substitution of natural gas for coal would reduce emissions of nitrogen oxides, sulfur dioxide, and particulates, it could increase hydrocarbons. However, the results of Model I indicate that the additional hydrocarbons would not significantly lessen the desirability of natural gas for air pollution control. This becomes evident when the emission reductions are valued at the respective shadow prices. For example, the emission *reductions* per unit of activity 27e have a total value of

$$(2.992) \ (\$.00428) + (-7.967) \ (\$.02476) + (15.599) \ (\$.32639)$$
$$+ (117.792) \ (\$.02193) + (14.63) \ (\$.07748) \cong \$8.62. \tag{3.5}$$

The fact that this value is positive indicates that the increased flow of hydrocarbons is offset by the decreased flows of the remaining pollutants.

In evaluating the desirability of converting coal furnaces to natural gas, there are numerous questions that relate to pollutant trade-offs. It is the capacity to account for complex trade-offs that makes linear programming an indispensable tool of analysis for air quality management.

The incremental costs in table 3.4 include $3.6 million to cover the higher cost of burning natural gas in place of coal in commercial, industrial, and residential furnaces. This represents an annual total of 14 billion cubic feet of natural gas for air pollution control, in comparison with the normal consumption of 284 billion cubic feet in 1975 (see sources 45, 46, 47, and 48 in table 2.1). The problem arises that this additional demand could force up the price of natural gas, especially if other airsheds are putting similar pressures

Table 3.4

Incremental costs of abatement

Alternative activities	Types of control	Total cost (dollars)
3b, 3c, 3d	Emission and evaporative control devices for new automobiles	16,536,917
11b	Low sulfur restriction on number six fuel oil	99,250
17e, 18e, 19e, 26e, 27e, 28e, 29e, 30e, 31e 32c, 37e	Conversion of industrial and commercial furnaces from coal to natural gas	2,251,451
19d, 20b, 20d, 21d, 22b, 23b, 36b	Upgraded efficiency for mechanical collectors and electrostatic precipitators for industrial and commercial furnaces	687,205
19d, 20d, 21d	Low sulfur coal for industrial and commerical furnaces	476,712
24b	Flue gas desulfurization system for industrial furnaces	359,700
25b	Substitution of coke oven gas for high sulfur coal in an industrial furnace	− 12,180
33b, 34b	Conversion of residential furnaces from coal to natural gas	1,388,720
38d, 41c 42c, 43b	Flue gas desulfurization systems for electric power plants	7,945,603
39b	Upgraded electrostatic precipitators for electric power plant	100,000
44b	Incremental cost of mine mouth generation of electricity	2,070,000
56b, 57b	Landfill disposal of wastes previously burned on site or at open dumps	1,746,626
59b, 60b	Municipal collection of leaves less value of leaves retained as compost	− 10,540
68b, 70b 71b, 86e	Fabric filters and secondary electrostatic precipitators for steel mills, wet scrubbers for blast furnaces, and dust control for coke charging	699,075
73b, 75b, 76b, 77b, 78b, 79b, 90b	Carbon monoxide waste heat boilers for petroleum refinery, wet scrubbers for fertilizer mixing operation, catalytic combuster for nitric acid plant, double catalytic process for sulfuric acid plants, and dust control for rock crushing	1,317,839

Alternative activities	Types of control	Total cost (dollars)
87b, 88b,	Recovery of coke oven gas and sulfuric acid (from lead smelting) for resale	−380,035
96b, 97b	Floating roofs for gasoline storage tanks	60,940
	Total	35,337,283

on coal burning. It is a limitation of the Linear Programming Model, as implemented here, that current prices are used to estimate control costs, without consideration of the pressures of increased demand on these prices. An examination of the data in table 3.4 suggests that other control methods, for which costs could be forced up because of the air pollution control needs, are flue gas desulfurization systems and landfill disposal of solid wastes. The model builder can correct this problem by selectively increasing unit costs in subsequent runs.

Sensitivity Analysis

The sensitivity of the solution of the Linear Programming Model to individual parameters can be tested. We shall examine the conditions for including a new activity in the solution and for removing an activity from the solution.

A process with a nonzero activity level, say 1a, has a *reduced cost*, ρ_{1a}, equal to zero; that is,

$$\rho_{1a} = C_{1a} - \sum_{i=1}^{6} e_{1a}^i \pi^i - \pi_1 = 0. \tag{3.6}$$

An inactive process, 1b, for that same source has a reduced cost that is greater than zero. This process would become active if C_{1b} declined by an amount greater than or equal to the reduced cost. The reader may confirm that process 1b (see table 2.1), the retrofit device for 1967 and earlier model vehicles, would be efficient if the unit cost were $11.73 instead of $26.30. Thus the reduced cost for process 1b is $26.30 minus $11.73, or $14.57. The reduced cost may be given the following economic interpretation. If an inefficient process j were utilized for a particular source, the unit saving in

Solution of the Empirical Model

abatement costs in the airshed that could be obtained by shifting to the efficient process for that source would be ρ_j. The vector of reduced costs corresponding to the optimal solution is part of the MPS/360 program output (see the appendix to this book).

There is a related test for the increase in cost for which an active process would become inefficient. Consider, for example, stack gas desulfurization process 42c in table 2.1. If the cost of this process increased sufficiently, it would become inactive, and process 42a or 42b would take its place. The reduced costs of these three processes in the solution of Model I are as follows:

$$
\begin{align*}
\rho_{42a} &= \$ .76, \\
\rho_{42b} &= \$1.30, \\
\rho_{42c} &= \$ .00.
\end{align*}
\tag{3.7}
$$

It follows that if C_{42c} were \$.76 more than the estimated cost, process 42a would become active. Thus the analysis of reduced costs permits the model builder to test the sensitivity of the solution $\mathbf{x}^*$ to his estimates of unit costs.

It is important in some cases to test the sensitivity of the model to a set of costs. For example, the solution of the St. Louis model indicated that 14 billion cubic feet of natural gas should replace approximately 850,000 tons of coal in industrial and residential stokers.[7] This compares to a projected total combustion of 2,100,000 tons of coal by these same sources in the year 1975, in the absence of pollution control regulations. To account for the relative scarcity of natural gas in comparison to coal, the cost of natural gas in the model was taken as the market price plus 18¢ per thousand cubic feet. The latter was an estimate of the increased cost of producing synthetic gas from coal (see the discussion in chapter 2, and in Kohn (1969a, pp. 562–569). However, the 18¢ figure would be either too low, if the cost of manufacturing gas were much higher, or too high, for if the 18¢ estimate is correct, it should have been discounted to the present value inasmuch as it represents a future cost burden. Accordingly, the model was rerun changing the cost of activities 17e, 18e, 19e, and all other activities that represent the use of natural gas for air pollution control. It was found that the optimal solution, $\mathbf{x}_j^*$, and the quantity of natural gas required, changed insignificantly even when the scarcity premium was doubled to 36¢. Only when the scarcity premium was reduced from 18¢ to 2¢ were there any significant increases in the extent of conversion to natural gas. Thus the solution of the model was relatively insensitive to this particular cost parameter.

It is also possible to test the sensitivity of the linear programming solution to emission factors. In the case of process 1b, in which

$$p_{1b} = C_{1b} - \sum_{i=1}^{6} e_{1b}^i \pi^i - \pi_1 \cong \$14.57, \tag{3.8}$$

the emission factor for, say, nitrogen oxides would have to change by $(\$14.57)/(-\$.32639)$, or approximately 45 pounds, for this process to be efficient at its current unit cost. The usefulness of the Linear Programming Model is greatly enhanced by this capability for testing the sensitivity of the solution to estimates of individual parameters.

Equiproportional Abatement

Kneese and Schultze (1975) suggest that regulatory agencies tend to require each source to reduce emissions by the same proportion so as to distribute the burden of pollution control equitably. This approach to pollution control can be illustrated with sources 68 and 72, whose individual and combined emissions prior to regulation are as follows:

$x_{68a} = 1,000,000$ tons of steel,
$x_{72a} = 2,400,000$ tons of grain,

$$\begin{aligned}
(10.6)x_{68a} &= 10,600,000 \text{ pounds of particulates}, &\qquad (3.9)\\
(6.0)x_{72a} &= 14,400,000 \text{ pounds of particulates},\\
\text{Total} &= 25,000,000 \text{ pounds of particulates}.
\end{aligned}$$

In the least-cost solution, $\mathbf{x}_1^*$, the emissions, in pounds of particulates, are

$$\begin{aligned}
(.16)x_{68b} &= 160,000,\\
(6.0)x_{72a} &= 14,400,000, &\qquad (3.10)\\
\text{Total emissions} &= 14,560,000.
\end{aligned}$$

The allowable emissions from the two sources combined constitute 58.24 percent of the preregulatory emissions, and the individual and combined costs of abatement (in dollars) are

$$\begin{aligned}
(\$.31)x_{68b} &= 310,000,\\
(\$.04)x_{72a} &= 96,000, &\qquad (3.11)\\
\text{Total cost} &= 406,000.
\end{aligned}$$

If these two sources were required to abate equiproportionally, the same level of emissions would be achieved with activity levels and corresponding total costs as follows:

x_{68a} = 576,000,
x_{68b} = 424,000,
x_{72a} = 1,316,497,
x_{72b} = 1,083,503,

$(10.6)x_{68a}$ = 6,105,600 pounds of particulates,
$(.16)x_{68b}$ = 67,840 pounds of particulates,
$(6.0)x_{72a}$ = 7,898,984 pounds of particulates, $\qquad$ (3.12)
$(.45)x_{72b}$ = 487,576 pounds of particulates,
Total $\quad$ = 14,560,000 pounds of particulates,

$(\$.00)x_{68a}$ = 0 dollars per year,
$(\$.31)x_{68b}$ = 21,030 dollars per year,
$(\$.04)x_{72a}$ = 52,660 dollars per year,
$(\$.49)x_{72b}$ = 530,916 dollars per year,
Total $\quad$ = 604,606 dollars per year.

As a group, the two sources would expend substantially less in the least-cost solution to obtain the same combined allowable flow of particulates.

The projected emission flows in the St. Louis airshed, assuming implementation of the least-cost set of abatement activities, are shown in table 3.5. A comparison of these data with the emission flows in table 2.2 indicates that the emission abatement is clearly not equiproportional. For example, 99 percent of the sulfur dioxide that would originate in the combustion of coal by residential users is eliminated in the least-cost solution, whereas the sulfur dioxide emissions from the combustion of fuel oil are reduced by only 10 percent.

Legal Regulations in the St. Louis Airshed Compared to the Least-Cost Activities

In the solution of Model I, no controls are required for municipal and privately owned incinerators (sources 52, 53, 54, 55), and neither iron and steel foundries (sources 63, 64, 65, 66) nor cement plants (sources 80, 81, 82) are required to increase their control efficiencies. This contrasts with present stringent regulations for such sources. It may be that control of these sources

Table 3.5
Emission flows in the St. Louis airshed in 1975 under the least-cost set of air pollution control regulations (millions of pounds[a])

Category of source	Carbon monoxide	Hydro-carbons	Nitro-gen oxides	Sulfur dioxide	Partic-ulates	Benzo-(a)pyrene
Transportation	2,247	495	93	14	26	338
Combustion of fuel oil	(less than 0.5)	1	33	50	4	8
Combustion of coal by industrial and commercial users	6	4	25	120	4	61
Combustion of coal by residential users	(less than 0.5)	3	1	(less than 0.5)	(less than 0.5)	1
Combustion of coal by public utilities	3	1	90	98	3	13
Combustion of natural and by-product gases	(less than 0.5)	125	57	45	6	22
Combustion of refuse	41	122	1	1	28	129
Industrial processes	38	51	4	72	56	0
Evaporation and miscellaneous minor sources	0	192	0	0	9	0
Total	2,335	994	304	400	136	572

[a]Benzo(a)pyrene emissions are in thousands of grams.

is not cost-effective in the present model because the pollution requirements are based on total annual emissions and ignore the neighborhood effects for which incinerators, foundries, and cement plants are noted. On the other hand, the Burton, Pechan, and Sanjour model, which *is* sensitive to neighborhood effects, yields solutions in which a number of point sources which are required to adopt additional controls under government regulations are left in their original state. (See Burton et al., 1973, p.415.) It appears to be the case that regulatory agencies compromise least-cost considerations with a desire to distribute the burden of abatement equitably among sources; some abatement is required of all sources, regardless of cost.

Economists have suggested that emission flows could be reduced at substantially lower cost than is required under typical regulations. Atkinson and Lewis, whose ELC model is included in table 3.2, estimate a 50 percent saving for a least-cost over a typical legal solution.

The potential saving for the St. Louis airshed in 1975 was estimated with the Linear Programming Model. From the variables listed in table 2.1, the set of activity levels x^r that most closely approximated the air pollution control regulations in the St. Louis region as of 1972 was selected. These represent a total cost of abatement, $\mathbf{C}x^r$, of $56.5 million. However, the emission flows $\mathbf{e}x^r$ corresponding to this solution differ from the allowable flows given in (2.12). They are somewhat higher for hydrocarbons, sulfur dioxide, and particulates, and lower for carbon monoxide and nitrogen oxides. Accordingly, Model I was rerun with requirements $\mathbf{e}x^r$ in place of (2.12). The resulting total cost of abatement was $44.8 million. This suggests that the least-cost solution achieves a saving of approximately 20 percent over the legal solution. If the base-year, nondiscretionary cost ($11.2 million) is deducted from the two costs, a saving of 25 percent for the least-cost solution is obtained, which is less than that predicted by Atkinson and Lewis. This is due in part to the fact that the same total emission flows $\mathbf{e}x^r$ are achieved here with the two sets of activity levels that are being compared, whereas Atkinson and Lewis compare a least-cost solution to a regulatory solution that achieves substantially better air quality. Furthermore, Anderson and Crocker (1971) have observed that Kohn's model yields a solution that is closer to the legal solution than those obtained from the models of other investigators. This could also explain why the savings achieved with the least-cost solution, as compared to the legal solution, are less for the present model than for the ELC model of Atkinson and Lewis.

Alternate Waste Flows

Ayres and Kneese (1969) have argued that gaseous, solid, and liquid wastes are interdependent, and that the unilateral control of one form of waste may simply convert it into alternate pollutant forms that are equally (if not more) noxious. The conclusion that follows is that a model which fails to account for all waste streams may lead to serious error.[8] A related problem that concerned the regulatory officials in St. Louis county was that the flue gas desulfurization process being tested at Meramec Power Plant (see activity 41c in table 2.1) would contribute salts to the waste water effluent.

The present model was extended to test the significance of the Ayres and Kneese argument with respect to air quality control. Three alternative waste forms were considered: liquid waste discharged to rivers, solid waste buried in landfills, and waste heat added to rivers. Emission factors for these wastes (see table 3.6 for examples) were assigned to each of the 250 activities in the Linear Programming Model. The extended model is referred to as Model II. It was found that the incremental flows of liquid and thermal wastes associated with the optimal solution x^*_1 (see column 11 of table 2.1) were quite small. However, the incremental flow of solid waste, that is, the recovered particulate matter and the refuse that would no longer be burned on-site, totalled 450,000 tons, representing a 50 percent increase in landfill tonnage for the St. Louis region. This result is consistent with the contention of Ayres and Kneese that air pollution control can significantly augment other waste streams. When the Linear Programming Model was run with an additional requirement that air pollution abatement beyond the base-year level not add to alternate waste streams, the total cost of abatement increased from \$46.5 to \$67.3 million.[9]

Rather than preclude any trade-offs between waste forms, it may be efficient to convert wastes to forms that are less costly to control. In Kohn (1971b) it was reasoned that liquid wastes in the St. Louis airshed could be purified for \$.25 per thousand gallons, solid waste recycled for \$1.40 a ton, and thermal waste harmlessly discharged through cooling towers at \$.04 per million Btu.

When Model II was run with these prices on the external wastes, the total purification and reprocessing cost associated with the by-product waste streams was \$500,000 per year. Although this is a substantial sum, it is less than 2 percent of the incremental costs of air pollution control. Because the

Table 3.6

Emission factors for liquid, solid, and thermal wastes for selected activities in Model II

Variable number	Activity unit and description
17a	Tons of coal burned by industry in unequipped underfeed stokers
17b	As 17a, with 85 percent efficient mechanical collectors
17e	As 17a, except replaced by natural gas
52a	Tons of municipal refuse burned in city incinerators
52b	As 52a, with wet scrubbers
56a	Tons of waste burned on-site
56b	As 56a, except compacted for strip mine landfill

[a]Bottom ash.

[b]Waste heat at the powerplant in generating electricity (.8 kilowatt hours) for operating the stoker.

[c]Bottom ash and collected fly ash.

[d]Waste heat at power plant associated with electrical requirements (6.7 kilowatt hours) for operating the blowers, plus the electrical requirement for the stoker.

[e]Grate residue.

[f]Waste heat at power plant associated with electrical requirements for operating the scrubbers.

Gallons of liquid waste	Pounds of solid waste	Btu of waste heat
0	170.[a]	3,680.[b]
0	212.5[c]	34,500.[d]
0	0	0
0	470.[e]	0
430.	470.[e]	18,400.[f]
0	700.[a]	0
0	2000.	0

alternate waste outputs were penalized in this model, a set of activities was chosen, which minimized the sum of such penalties plus the costs of air pollution control. As a result the least-cost set of air pollution abatement activities increased by $60,000 a year. This was only a slight increase, but it was instructive; two of the activities that became nonzero in the revised model are activities that recover *and recycle* particulate matter. These are 72b (grain) and 81b (cement). To the extent that the original model ignores real but unpriced benefits of recycling recovered pollutants, the solution is not efficient. In the case of the present model, however, any such errors do not appear to be as serious as Ayres and Kneese anticipated.

The Least-Cost Solution as a Regulatory Guide: The Importance of Enforcement Costs

The control method solution indicates how total pollution flows can be reduced to allowable levels at least cost. This can provide useful information for designing pollution control regulations and emission standards. For example, the fact that activity 70b is optimal may suggest that steel producers be required to use both primary (cyclone) and secondary (electrostatic precipitator) controls for their sinter production. Alternatively, an emmision rate of $e_{70b} = .2$ pounds of particulates per ton of sinter could be established. However, it is unlikely that the entire solution x_i^*, should be translated into a set of legal standards. A major reason for this conclusion is that the unit cost coefficients in table 2.1 do not include administrative and enforcement costs that would be incurred by government agencies. Not only is the total cost of abatement thereby understated,[10] but it is not clear that the same solution $\mathbf{x}_1^*$ would be optimal. It is likely that a set of regulations that required certain furnaces to be switched to low sulfur coal and others to natural gas, and allowed some to continue uncontrolled, would be difficult to administer. If enforcement costs were properly assigned to individual control activities in the model, the efficient solution would probably indicate a more uniform treatment of comparable sources.[11]

The significance of administrative and enforcement costs is accounted for in the Siegel, Ehrenfeld, and Morganstern model (see table 3.2). The decision variables in this model are not independent abatement activities but *strategies*. These are clusters of activities that apply uniformly to comparable sources and would therefore be more readily enforceable.

It would be difficult to revise the data in table 2.1 to include enforcement costs. The set of feasible activities $\mathbf{x}$ would have to be increased enormously to account for different enforcement costs under different combinations of activities. If the Linear Programming Model is intended to generate a set of legal regulations, a model more like that of Siegel, Ehrenfeld, and Morganstern would be desirable. (The present formulation of the model, in which governmental costs are excluded, has the advantage that the optimal solution would, in theory, be voluntarily instituted in a program in which Pigouvian fees, equivalent to the pollutant shadow prices, are imposed on polluters. This will be discussed at length in chapter 7.)

There may also be considerations of equity that would cause a regulatory agency to deviate from the least-cost solution; these will be discussed in chapter 7. The major usefulness of the Linear Programming Model for regulatory agencies lies in its capacity for evaluating the multidimensional pollutant and cost trade-offs that are so important in air quality management.

Multiple Optima and Divisibilities in the Control Method Solution

The Control Method solution $\mathbf{x}_1^*$, as given in column 11 of table 2.1, is not unique. There are alternate abatement activities for sources 19, 20, and 21 that achieve the same set of emission flows at the same total annual cost.

The reader may confirm that the following pairs of activity levels are equivalent:

$$\begin{bmatrix} x_{19b} = 0 \\ x_{19d} = 66{,}970 \\ x_{19e} = 106{,}030 \\ x_{20b} = 142{,}000 \\ x_{20d} = 216{,}800 \\ x_{21b} = 0 \\ x_{21d} = 97{,}600 \end{bmatrix}, \quad \begin{bmatrix} x_{19b} = 44{,}400 \\ x_{19d} = 22{,}570 \\ x_{19e} = 106{,}030 \\ x_{20b} = 0 \\ x_{20d} = 358{,}800 \\ x_{21b} = 97{,}600 \\ x_{21d} = 0 \end{bmatrix}. \tag{3.13}$$

Both represent \$1,798,751 in abatement costs; 381,370 tons of low sulfur coal; 39,915,744 pounds of sulfur dioxide emissions, etc. There are, in fact, an infinite number of optimal alternative combinations of the above activities.

The activity sets in (3.13) illustrate the case in which two or more activity levels for a single pollution source are nonzero. This is referred to as a *divisibility* in the solution. There is some controversy as to whether divisibilities, which imply two types of control for a single source, lessen the applicability of a linear programming model to regulatory policy making. Often it is possible to disaggregate the affected source so as to justify separate control activities. On the other hand, the problem of divisibility can be avoided with integer programming. For any source i that cannot be subdivided, it would be required that the control method activity level be either zero or s_i. In the case of the present model, the specification of an integer solution would increase the total cost of abatement and reduce emission flows below allowable levels for some if not all of the six pollutants.[12] In general, however, the number m of pollution sources will be much greater than the number p of pollutant requirements. The number of sources for which there is divisibility cannot exceed p; therefore the problem of divisibility, if indeed it is a problem, is limited to the fraction p/m of the pollution sources.

Verification

It is appropriate that an empirical model be subject to verification. Wherever possible in this book, this will be attempted. For example, it is useful to test whether the estimated total cost of abatement obtained with Model I is a realistic figure. The only substantial data available on which to make a comparison is a published estimate of \$16.8 million annual operating costs by industrial sources for air pollution abatement in the St. Louis SMSA in 1975.[13]

Of the total costs in table 3.1, approximately $9.4 million can be allocated to industrial sources. This outlay is in 1968 dollars. Based on the wholesale price index for industrial commodities, this is equivalent to $15.6 million in 1975 dollars. Because this represents a least-cost solution, it follows from the discussion on p. 92 that this sum would be less than the actual costs associated with a regulatory solution. This provides some support for the overall size of the numbers used in this study.

Although this comparison of total industrial costs may be favorable, this does not confirm that individual components of the least-cost solution are all correct. It is now apparent that the data used to implement the Linear Programming Model was incomplete, and in places incorrect.[14] The relative costs for substitute fuels, flue gas desulfurization, and automobile controls are higher than had been anticipated.[15] Furthermore, some significant pollution sources had not yet been identified, and the aggregation of other sources, particularly industries involved in hydrocarbon processing (see source 89 in table 2.1), resulted in an understatement of control cost. Fortunately, the implementation of such models today will benefit by the availability of much better emission, technological, and cost data.

Despite the limitations of the data, the usefulness of the model has been demonstrated. It can be used to resolve difficult policy problems, particularly when there are interdependencies in the pollutants being controlled. This was shown in this chapter in the evaluation of the low sulfur coal controversy, the assessment of the desirability of converting coal to natural gas despite the relative scarcity of the latter and the possibility of increased leakage, and the perspective gained on the practical significance of alternative joint-waste flows. Furthermore, it has been shown that where the reliability of certain data is in doubt; sensitivity analysis may indicate a substantial range of values for the uncertain data over which the solution of the model does not vary significantly. This adds enormous power to the Linear Programming Model and enhances its usefulness as a potential guide for regulatory decision making.

4

EMISSION FLOWS
AND POLLUTANT
CONCENTRATIONS

The diffusion of emissions in the ambient air results in pollutant concentrations. These are ratios, such as micrograms of a pollutant per cubic meter of air (μg/m³) and cubic meters of a gaseous pollutant per million cubic meters of air, commonly called parts per million (ppm). It is the concentrations in the ambient air that affect people and their environment.

In this chapter we shall examine the relationship between emission flows and pollutant concentrations. A simple theory is one in which an airshed is viewed as a shallow box. Some preliminary results based on the Box formula provide the basis for a linear relationship between total emissions and ambient air concentrations. This linear relationship is utilized in the Larsen formula and in the Stochastic formula. A more sophisticated relationship between emission flows and concentrations is defined by the Diffusion formula. In that relationship, the impact of a pound of pollution on an ambient air concentration depends upon the location of the emitting source. In versions of the Linear Programming Model that incorporate the Diffusion formula, pollution control can be achieved by strategic location of sources as well as by technological abatement. The chapter concludes with a comparison of economic models based on the alternative relationships between emission flows and pollutant concentrations.

The Box Formula

An airshed may be viewed as a box in which emissions of each pollutant diffuse completely and there is a uniform concentration of that pollutant throughout the entire box. Smith (1961) and Wanta (1968) have developed a box concept that yields a proportional relationship between total emissions and concentrations. According to their theory, the concentration, q^i equals the flow of the pollutant f^i divided by the volume of air in which it is mixed. Taking the area of the city as a square with side, s, vertical mixing height h, and annual wind speed v, the rectangular volume of air is approximately vhs; and the equilibrium annual average concentration is

$$q^i = f^i/(vhs). \tag{4.1}$$

The Box formula may be illustrated for sulfur dioxide, using data for the

St. Louis airshed in 1970. In that year an estimated 1,375 million pounds of sulfur dioxide were emitted into the airshed (see Kohn and Weger, 1973). Based on an average annual wind speed of 15,400 meters per hour, a mixing height of 1210 meters, and a width of 100,000 meters for the airshed, the annual average concentration of sulfur dioxide would be[1]

$$q^i = \frac{(1,375,000,000)\ (454)\ (1,000,000)}{(15,400)\ (8760)\ (1,210)\ (100,000)} \cong 38\ \mu g/m^3. \tag{4.2}$$

This is equivalent to a concentration of .014 parts per million, which is approximately one half of the 1970 annual average concentration for that pollutant as observed at the CAMP (Continuous Air Monitoring Program) station.[2] The concentrations of the remaining pollutants for the year 1970 (as predicted from formula (4.1), using the above data for v, h, and s and estimated emission flows for 1970) are listed in column 3 of table 4.1. The observed concentrations at the CAMP station are listed in column 7. Inasmuch as the CAMP station concentrations are generally much higher than concentrations elsewhere in the airshed, we would expect them to be higher than the averages predicted with the Box formula. In view of the simplistic nature of the latter and the fact that complete mixing is assumed, the predictions derived are surprisingly good. They do suggest that a linear relationship between concentrations and total emissions may be taken as a first approximation. The general linear relationship is

$$q^i = mf^i + b^i, \tag{4.3}$$

where b^i is the background concentration of the ith pollutant. According to the Box formula, the value of m (for calculating concentrations in grams per cubic meter) would be the same for each pollutant, that is $m = 1/(vhs)$.

The Larsen Formula

Larsen (1961) has suggested that a given percentage reduction in the total annual emission flow, f^i will result in the same percentage reduction of the net pollutant concentration, $(q^i - b^i)$. This implies a relationship of the form,

$$q^i = m^i f^i + b^i. \tag{4.4}$$

Like the Box formula, the relationship is linear. However, uniform mixing is no longer assumed and the value of m^i, which is both pollutant and location specific, is determined statistically. In the most simple application, the Larsen constants m^i are determined from known data from some previous, single year. For example, the particulate concentration near the St. Louis CAMP station in 1963 was 128.3 $\mu g/m^3$, as compared to an estimated background concentration of 31 $\mu g/m^3$. The total annual flow of particulate emissions in 1963 was approximately 300 million pounds. Therefore, the Larsen constant for particulates at the CAMP station is

$$m^i = (128.3 - 31.0)/(300,000,000) = .324 \times 10^{-6}. \tag{4.5}$$

The Larsen constants for all of the pollutants are included in table 4.2. Each was derived from observed data for the year 1963. These constants and the emission flows in column 2 of table 4.1 were used to calculate the CAMP station concentrations for 1970. The resulting estimates are shown in column 5 of table 4.1 and may be compared with the observed concentrations in column 7.

With the exception of particulates, the Larsen formula overestimated the 1970 CAMP station concentrations (see table 4.1). The overestimate was most serious for sulfur dioxide. In 1963 annual emissions of that pollutant in the St. Louis airshed totalled approximately 1,178 million pounds, and the measured concentration of sulfur dioxide at the CAMP station was .059 ppm (see Kohn and Weger, 1973). By 1970 the emissions of sulfur dioxide had increased to 1,375 million pounds. Based on the Larsen formula, the concentration at the CAMP station in that year should have been .070 ppm, yet it had actually declined to .030 ppm. The Larsen formula failed to account for the fact that there was a relatively large decrease in the combustion of high sulfur coal by sources near the CAMP station and for the fact that the increase in total emissions of sulfur dioxide occurred in new power plants located farther and farther away from the CAMP station in the years following 1963, the year for which data were taken to compute the Larsen constants.

Because emissions do not mix evenly throughout an airshed, the geographic location of pollution sources in relation to the receptor station is a significant factor affecting the pollutant concentrations. To the extent that the locational pattern of emissions changes from what it had been in the base-year, the Larsen formula becomes less reliable.[3]

Table 4.1

Calculated and observed annual average pollutant concentrations at the St. Louis CAMP station in 1970

(1) Pollutant	(2) Emissions in 1970 (millions of pounds)[a]	(3) Box formula
Carbon monoxide	3,109	.1 ppm
Hydrocarbons[b]	939	1.5 ppm
Nitrogen oxides	420	.007 ppm
Sulfur dioxide	1,375	.014 ppm
Particulates[c]	233	37 $\mu g/m^3$

[a]These flows were estimated for research that is partially published in Kohn and Weger (1963).

[b]Background concentration is 1.5 ppm. Subsequent data suggest that this estimate may be high.

[c]Background concentration is 31 $\mu g/m^3$.

[d]Each concentration in column 6 is obtained by multiplying the corresponding concentration in column 5 by the ratio of the concentration for that pollutant in 1963, as predicted with the Diffusion formula, to the actual observed concentration in 1963.

[e]The 1970 CAMP station measurements were computed from data provided by Will Hager, Division of Air Pollution Control, City of St. Louis (see Kohn and Weger, 1973).

(4) Larsen formula	(5) Diffusion formula	(6) Calibrated Diffusion formula[d]	(7) Observed concentrations[e]
6.7 ppm	1.5 ppm	6.8 ppm	3.7 ppm
3.0 ppm	2.1 ppm	3.1 ppm	1.7 ppm
.095 ppm	.068 ppm	.082 ppm	.077 ppm
.070 ppm	.036 ppm	.037 ppm	.030 ppm
105 $\mu g/m^3$	68 $\mu g/m^3$	100 $\mu g/m^3$	130 $\mu g/m^3$

Determining Total Allowable Annual Emission Flows with the Larsen Formula

The allowable annual flows in Model I, which are listed in (2.12), were derived from the Larsen constants. This will be illustrated for particulate emissions. Given an air quality standard of 75 $\mu g/m^3$ for particulates, it follows that the total annual allowable flow of particulate emissions in the airshed, $\hat{f}^i$, is

$$\hat{f}^i = (\hat{q}^i - b^i)/m^i = (75 - 31)/(.324 \times 10^{-6}) \cong 135{,}800{,}000 \text{ pounds.} \quad (4.6)$$

Table 4.2 contains the air quality standards $\hat{q}^i$ and the corresponding allowable emission flows $\hat{f}^i$ that were calculated with the Larsen formula. The standards for sulfur dioxide and particulates are the legal standards in Missouri. There were no annual standards for the remaining pollutants, and, therefore, arbitrary concentrations were selected for carbon monoxide, hydrocarbons, and nitrogen oxides.

The Larsen formula, in matrix notation, is

$$\mathbf{q} = \mathbf{mf} + \mathbf{b}, \quad (4.7)$$

where $\mathbf{m}$ is a $p \times p$ diagonal matrix whose nonzero element is m^i. The pollution constraint in Model I is based on the relationship

$$\mathbf{f} = \mathbf{ex}. \quad (4.8)$$

Table 4.2

Larsen constants, background levels, standards, and allowable emission flows

Pollutant	m^i Larsen constant (concentration per pound of pollution)[a, b]	b^i Background concentration[c]	$\hat{q}^i$ Air quality standard (annual average concentration)	$\hat{f}^i$ Allowable annual emission flow (millions of pounds)
Carbon monoxide	$.214 \times 10^{-8}$	0	**5 ppm**	2335
Hydrocarbons	$.161 \times 10^{-8}$	1.5 ppm	3.1 ppm	994
Nitrogen oxides	$.227 \times 10^{-9}$	0	.069 ppm	304
Sulfur dioxide	$.050 \times 10^{-9}$	0	.02 ppm	400
Particulates	$.324 \times 10^{-6}$	31 $\mu g/m^3$	75 $\mu g/m^3$	136

[a]The Larsen constants and the allowable annual flows are rounded out for this table.

[b]The Larsen constant for benzo(a)pyrene is $(.158)$ (10^{-14}) $\mu g/m^3$ per microgram of emissions. The allowable flow of this pollutant in equations (2.12) is based on a background level of zero and a standard of .001 $\mu g/m^3$.

[c]The references for the background concentrations are contained in Kohn (1971d, p. 987). The following background concentrations, taken from *Cleaning Our Environment: The Chemical Basis for Action*, American Chemical Society, Washington, D.C., 1969, p. 24, were small and were ignored. They are .1 ppm for carbon monoxide, .001 ppm for nitrogen dioxide, and .0002 ppm for sulfur dioxide.

It follows, by substitution, that an equivalent version of Model I, which we shall call Model III, would contain the following constraint:

$$\mathbf{mex} \leqslant \hat{\mathbf{q}} - \mathbf{b}. \tag{4.9}$$

This formulation permits the air quality constraints to be directly expressed in concentrations rather than emission flows.

Other empirical models using a linear or statistical relationship between total emission flows and pollutant concentrations are described in table 4.3. The San Diego Clean Air Project is based on 1970 data relating total emissions of reactive hydrocarbons for that year to maximum *one-hour* concentrations of oxidants. The Linear Programming Model as applied to the St. Louis airshed is based on *annual* average concentrations, which are presumed to be the best and most reliable measure of air quality (see Snee and Pierrard, 1977).[4]

The Trijonis model (see table 4.3) uses statistical analysis to estimate a relationship between total annual emissions of reactive hydrocarbons and of nitrogen oxides in Los Angeles County, and the percentage of days per year in which specific one-hour pollutant concentrations are exceeded. The analysis, based on observations for a number of years rather than some single base-year, resulted in nonlinear relationships. In addition, Trijonis was able to relate oxidant concentrations to emissions of both precursor pollutants.

A Stochastic Formula

The Box formula and the Larsen formula imply deterministic relationships between concentrations and total emission flows. Van Belle and Schneiderman (1973, p. 322) note that a number of deterministic environmental models have been made stochastic by assuming that one of the variables in the model is random. The single meteorological variable in the Box formula,

$$q^i = (1/vhs)f^i + b^i, \tag{4.10}$$

is average wind speed v. This formula can be made location and pollutant specific without embedding a fixed value for v, as follows:

$$q^i = (d^i/v)f^i + b^i. \tag{4.11}$$

A stochastic version of the Linear Programming Model, called Model IV, was set up, in which the d^i were derived from the m^i using the relationship

$$m^i = d^i/v \tag{4.12}$$

Table 4.3

Empirical models of air quality management with linear and statistical relationship of total emissions to concentrations

Model	San Diego Clean Air Project	Trijonis
Empirical application	San Diego County (1975)	Los Angeles County (1975)
Air quality indicator	Maximum one-hour concentration for oxidants	Expected number of days per year in which one-hour concentrations of reactive hydrocarbons, nitrogen oxides, and ozone exceed California standards
Receptor points	Six	One
Concentration/emission relationship	Proportional relationship between oxidant concentration and total emissions of reactive hydrocarbons	Statistical, relating total emissions to concentrations
Objective	Minimize annual strategy expenditures	Minimize total cost of abatement
Types of sources	Fixed and mobile	Twenty-three types of point and moving sources
Abatement alternatives	Fixed source controls, retrofit devices for automobiles, inspection-maintenance, transportation management to reduce vehicle miles travelled	Thirty-one alternative control devices plus existing control methods
Reference	Bruce F. Goeller et al., *San Diego Clean Air Project: Summary Report*, Rand Corp., Santa Monica, 1973	John C. Trijonis, "Economic Air Pollution Control Model for Los Angeles County in 1975," *Environmental Science and Technology*, 8, Sept. 1974, pp. 811–826

and the annual average wind speed v of 9.6 miles per hour. This value of v is the mean annual average wind speed at Lambert Airport in St. Louis for the twelve years 1958–1969. If we assume that wind speed is an independent, normally distributed random variable, it follows that there is a 50 percent probability that the average annual wind speed will equal or exceed the mean annual average. This means that the solution $\mathbf{x}^*$ to the Linear Programming Model I or III will satisfy the air quality standards with a probability of only 50 percent. There is a 50 percent probability that v will be less than 9.6 miles per hour and that the air quality goals will not be met. Actually, because the mean value of v used in the calculations is only a statistical estimate of the true mean, it turns out that we can be only 50 percent confident that the probability is 50 percent, that is, that the true mean value of v equals or exceeds 9.6 miles per hour and that the solution $\mathbf{x}^*_{\text{III}}$ to the Linear Programming Model will satisfy the air quality standards.

There is a substantial cost penalty for greater certainty. This was demonstrated with Model IV, in which the air quality constraint is

$$\mathbf{ex} \leqslant \mathbf{d}^{-1}(\hat{\mathbf{q}} - \mathbf{b})v, \tag{4.13}$$

where $\mathbf{d}^{-1}$ is the inverse of the $p \times p$ diagonal matrix whose nonzero element is the constant d^i, and v is the average wind speed.[5] The following conclusions were obtained with Model IV. To be 50 percent confident of a 99 percent probability of achieving the air quality goals in the St. Louis airshed in 1975 would require a control strategy based on an annual average wind speed of 8.4 miles per hour rather than 9.6 miles per hour. It follows from (4.13) that the air quality constraints $(\hat{q}^i - b^i)$ would each decrease by the ratio $(8.4)/(9.6)$: from 5.0 ppm of carbon monoxide to approximately 4.4 ppm, from .020 ppm of sulfur dioxide to approximately .018 ppm, etc. As a result of tightening the constraints to allow for the possibility of the lower average annual wind speed, the total annual cost of abatement for 1975 would increase by over 50 percent, from \$46.5 million to \$71.3 million. Furthermore, to be 95 percent confident of a 99 percent probability of achieving the air quality standards would require a control strategy based upon a still lower value of v: 7.8 miles per hour. This would increase the total annual cost of pollution abatement in the St. Louis airshed by 90 percent.

Reducing the allowable emission flows to increase the probability of achieving the air quality standards is very costly. An alternative approach, suggested by Teller (1967), is to forecast periods of low wind speed and institute temporary control measures in such periods. What Teller calls "forecasting abatement" would include short term substitutions of low sulfur fuels and temporary curtailment of certain polluting activities. Teller estimated that the sulfur dioxide standards in the Nashville area could be achieved by forecasting abatement at one sixth the annual cost of constant abatement. Because of the high cost of setting allowable annual emission flows low enough to account for all adverse meteorological conditions, there may well be a place for episode control. It is generally believed, however, that the primary emphasis in pollution control should be on permanent abatement activities. Not only is it risky to depend on emergency measures,[6] but such a strategy is likely to result in borderline air quality most of the time. Alternatively, if allowable emission flows are based on an annual average wind speed set below the estimated average, there is not only a greater certainty of achieving the standards but a likelihood of enjoying air quality *superior* to those standards.

Diffusion Formula

According to atmospheric diffusion theory, the effect of emissions from a particular source on the pollutant concentration at the receptor station depends on the location of that source. Both of the economic models described in table 4.4 are based on diffusion formulas.

A simple formula for the contribution to the long run pollutant concentration by a source emitting at a constant rate has been developed by Turner (1969). The following is an adaptation of the Turner formula. The contribution m_j of a pound of pollution from the jth source to the concentration of that pollutant at the receptor station k_j meters downwind is

$$m_j = \left[\frac{(265)\,\theta_j}{v_j k_j^{1.915}} \right] \exp \left[\frac{-H_j^2}{.0242 k_j^{1.83}} \right] \mu g/m^3. \tag{4.14}$$

The constant 265 adapts the Turner formula to the present model in which emission flows are expressed in pounds per year. The variable H_j is the effective stack height, in meters, of the source.

The variables θ_j and v_j in (4.14) may be explained by illustration. The St. Louis airshed was arbitrarily subdivided into sixteen compass sectors as shown in figure 4.1. For a point source located at A, the distance in meters to the CAMP station is k_j; and θ_j is the fraction of the year that the wind direction is from the west-southwest. The greater this fraction, the larger will be

Table 4.4

Empirical models of air quality management based on diffusion formulas

Model	Atkinson and Lewis Ambient Least Cost (ALC) model	Implementation Planning Program (IPP)
Empirical application	St. Louis region	New York and Philadelphia
Air quality indicator	Annual average particulate concentration in micrograms per cubic meter of air	Micrograms of sulfur dioxide per cubic meter of air and micrograms of particulates per cubic meter of air, each expressed in averaging times ranging from one hour to one year
Receptor points	Nine	Enough receptor points (as many as 275) to generate pollution isopleths
Concentration/emission relationship	Diffusion model	Diffusion model
Objective	Minimize total cost of abatement	Minimize total cost of pollution abatement
Types of sources	Twenty-seven large point sources accounting for 80 percent of total particulate emissions. Emissions from all other sources including area sources were added to the background level.	All types, but major emphasis on point sources
Abatement alternatives	Upward rising (piecewise linear) marginal cost of abatement schedules for each source	Thirty-six to fifty pollution reduction devices or methods
Reference	Scott E. Atkinson and Donald H. Lewis, "A Cost-Effectiveness Analysis of Alternative Air Quality Control Strategies," *Journal of Environmental Economics and Management*, 1, November, 1974, pp. 237–250.	Ellison S. Burton, Edward H. Pechan, and William Sanjour, "Solving the Air Pollution Control Puzzle," *Environmental Science and Technology*, 7, May, 1973, pp. 412–415.

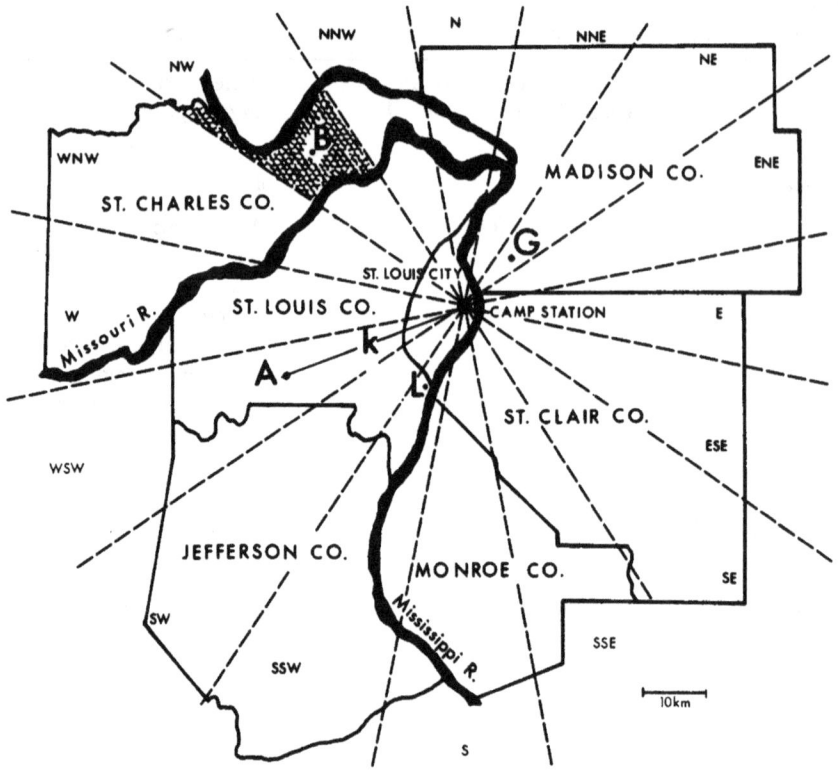

Figure 4.1
The St. Louis airshed subdivided into compass sectors

the contribution from the source at A to the concentration at the CAMP station. The value of v_j is the average annual wind velocity, in meters per second, through the sector in which the jth source is located.[7] Uniform horizontal dispersion along the path from point A to the CAMP station is assumed. A constant representing vertical dispersion for Class C Stability is merged into the above formula (see Kohn and Weger, 1973).

For ground level emissions $(H_j = 0)$, the concentration at the CAMP station is inversely proportional to $k^{1.915}$, which is approximately k^2. This is intuitively meaningful, as can be seen in figure 4.2, for the area of the semicircle DEF (over which pollution at distance k disperses) is proportional to k^2.

The lower the wind velocity v_j, the greater will be the accumulation of pollution at any point downwind from A for any averaging time. The greater the height of the stack, the lower the concentration, although the effect of stack height diminishes with distance.[8]

Formula (4.14) is designed for point sources. In the case of area sources (automobiles, residential furnaces, incinerators, etc.) the emissions are allocated to the compass subsector of the individual counties. There are forty-four such subsectors, one of which is indicated by the shaded area in figure 4.1. The emissions from an area source within a subsector are assumed to be evenly distributed throughout that sector (although for the sectors in the city of St. Louis it is assumed that such emissions begin at certain finite distances from the CAMP station). A methodology of integration was used to find a specific distance from the CAMP station such that if emissions from the area source in that subsector were concentrated at a point (for example, B in figure 4.1) at the calculated distance, the same concentration contribution would result, using formula (4.14), as would be derived by integrating concentration contributions from every point within the sector.

The calculation of a typical diffusion factor m_j may be illustrated for activity number 70 (see table 2.1), which represents the production of sinter at Granite City Steel Company. This company is located six miles (9,656 meters) northeast of the St. Louis CAMP station (see point G in figure 4.1). The proportion θ_{70} of winds from the northeast to all winds is .039, and the average velocity of winds from that direction is 3.53 meters per second. Assuming a stack height of 75 meters, as is conventionally done, m_{70} can be calculated with diffusion formula (4.14) as follows:

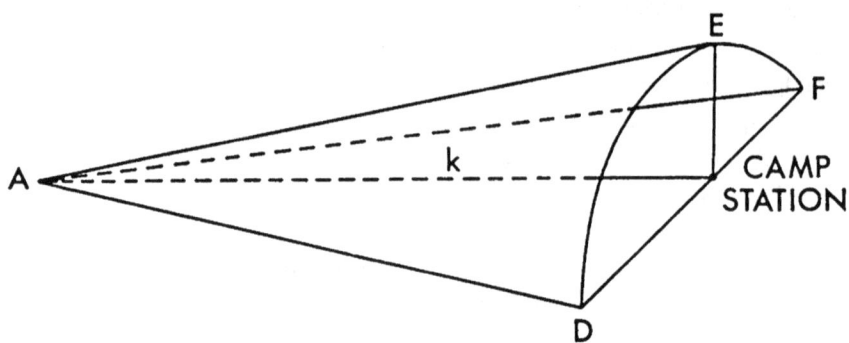

Figure 4.2
Downwind convection and horizontal and vertical dispersion of pollution

$$m_{70} = \left[\frac{(265)\ (.039)}{(3.53)(9656)^{1.915}}\right] \exp\left[\frac{-75^2}{(.0242)(9656^1)^{.83}}\right] \qquad (4.15)$$

$\cong 677 \times 10^{-10}\ \mu g/m^3$ per pound of pollutant.

This indicates that each pound of a pollutant emitted by activity 70a (and/or 70b) contributes 677×10^{-10} micrograms per cubic meter to the annual, arithmetic-average concentration of that pollutant at the St. Louis CAMP station.[9]

Estimating Concentrations with the Diffusion Formula

Emission flows and pollutant concentrations are related by the Diffusion formula in the matrix equation

$$\mathbf{q} = \mathbf{geMX} - \mathbf{b}, \qquad (4.16)$$

where $\mathbf{g}$ is a $p \times p$ diagonal matrix whose nonzero element, g^i, converts micrograms per cubic meter into parts per million for the gaseous pollutants.[10] The activity vector $\mathbf{X}$ in (4.16) is upper case to distinguish it from the activity vector $\mathbf{x}$. With the Diffusion formula there is a disaggregation of pollution sources according to their location in the airshed. Accordingly, there are M sources[11] rather than m, and N activity variables rather than n. This increases the number of pollution sources in the Linear Programming Model from 97 to 150 and number of activity variables from 250 to more than 500.

In (4.16) the $\mathbf{e}$ matrix is $p \times N$ and the $\mathbf{M}$ matrix is an $N \times N$ diagonal matrix whose nonzero element m_j converts pounds of each pollutant from source j into micrograms per cubic meter at the CAMP station, according to (4.14).

Formula (4.16) was used to estimate pollutant concentrations at the St. Louis CAMP station in 1963 and 1970. This was accomplished by substituting a vector $\mathbf{X}_t$ for $\mathbf{X}$, where $\mathbf{X}_t$ represents the actual activity levels in year t for each of the 500 activities. The estimated concentrations for 1970 are contained in column 5 of table 4.1. Because the estimates were less than the observed concentrations (see column 7), the Diffusion formula was "calibrated." The calibration was based on the observed and the predicted concentrations for the year 1963. Each element g^i in (4.16) was multiplied by the ratio of the observed 1963 net concentration to the calculated 1963 net concentration for that pollutant.[12]

The predicted concentrations obtained with the Calibrated Diffusion

Table 4.5

Emission abatement for 1975 in alternative versions of the Linear Programming Model

Pollutant	Standard	Larsen formula (Model III)	
		Predicted concentration before abatement[a]	Emission abatement (millions of pounds)
Carbon monoxide	5.0 ppm	9.0 ppm	1,867
Hydrocarbons	3.1 ppm	3.9 ppm	524
Nitrogen oxides	.069 ppm	.094 ppm	112
Sulfur dioxide	.02 ppm	.069 ppm	989
Particulates	75 $\mu g/m^3$	128 $\mu g/m^3$	164

[a]The predicted concentrations for 1975 may be calculated from the preregulatory emission flows in table 2.2 using the Larsen constants in table 4.2.
[b]These concentrations were predicted with formula (4.16) and $\mathbf{X_a}$, where $\mathbf{X_a}$ is the preregulatory vector analogous to $\mathbf{x_a}$.

formula are contained in column 6 of table 4.1. The calibration improved the prediction for particulates, correcting for the fact that the Diffusion formula does not take into account the effect of gravity on suspended particulate matter. That the Calibrated Diffusion formula overestimated carbon monoxide, hydrocarbons, and nitrogen oxides, which are for the most part automotive pollutants, may reflect the fact that between 1963 (the year on which the calibrations are based) and 1970, the completion of an expressway diverted a large volume of traffic that formerly flowed right past the CAMP station. Unfortunately, the methodology for handling this particular area source does not take into account such shifts in traffic patterns.

The Linear Programming Model Incorporating the Calibrated Diffusion Formula

Version V of the Linear Programming Model incorporates the Calibrated Diffusion formula and has the form

Minimize $\mathbf{z} = \mathbf{CX}$

subject to $\mathbf{uX} = \mathbf{s}$,

$$\mathbf{geMX} \leqslant \hat{\mathbf{g}} - \mathbf{b},$$

$$\mathbf{X} \geqslant 0.$$

(4.17)

Calibrated Diffusion formula (Model V)	
Predicted concentration before abatement[b]	Emission abatement (millions of pounds)
9.7 ppm	2,131
4.0 ppm	483
.094 ppm	46
.045 ppm	333
131 μg/m	146

Because of locational disaggregation, the dimensions of the **C**, **u**, **s**, and **e** matrices are larger than the corresponding matrices in Model I. The solution of Model V yields a total cost of abatement $\mathbf{CX}_V^*$ of $44.4 million for 1975.

This total cost is approximately $2.1 million less than the total cost for Model I (or III). It is useful to consider why the total cost of abatement might be less for Model V. Because the contribution of any source to the concentration at the CAMP station falls off rapidly with the distance of separation, it is the closer sources that have the major impact. As a consequence, it is likely (although not necessary) that a smaller quantity of emissions can be eliminated to achieve the required concentrations. This is reflected in table 4.5 where it can be seen that the total abatement of emission flows to achieve the air quality standards is less for four of the pollutants in Model V.

Offsetting these savings, however, was the fact that more costly processes had to be used. Because the major power plants, which are relatively less costly to control, are located from 27 to 57 kilometers away from the CAMP station, abatement by these particular sources has a small effect on the concentrations at the receptor.[13] Accordingly, the total cost of abatement for public utilities was only $5.3 million in Model V as compared to $15.3 million (see table 3.1) in Model III. By contrast, the control of transportation sources, industrial fuel combustion, refuse disposal, and industrial processes was relatively more intense because these represented activities closer to the CAMP station.[14]

In comparable research, Atkinson and Lewis (1974, p. 247) found that the total cost of abatement, derived from an economic model based on a diffusion formula (see table 4.4), is 50 percent less than the total cost with a model utilizing a linear relationship of total emissions to concentrations (see table 3.2). The implication in their work is that the Diffusion formula inevitably results in a lower total cost of abatement. Fisher and Peterson (1976, p. 25) have hailed the significance of their finding, and it has been given a theoretical basis by Goeller et al. (1973, p. 122), who state that the Diffusion formula, in contrast to a linear relationship such as they themselves employed (see table 3.2), may make it "possible to find strategies that exploit geography and meteorology to achieve the standards at somewhat lower cost."

It is conceivable, however, that the total cost of abatement would be *more* in a linear programming model incorporating a diffusion formula. This could be the case if the critical sources close to the receptor station were very costly to control. Even though a lesser reduction in emissions would suffice, this could require a larger total cost. It could also be the case, if the more critical sources could not be controlled, that a greater reduction in total emissions would be necessary. This, in fact, was the case for carbon monoxide in Model V (see table 4.5).

Pollution Control by Locational Selectivity

When the relationship of emissions and concentrations is defined by the Diffusion formula, pollution control can be achieved by selective location of sources as well as by emission abatement. Linear Programming Model V was used to illustrate the trade-off between pollution control by plant relocation and technological abatement. This was done by simulating alternative locations for Granite City Steel Company.[15] The polluting activities of this company are represented by source numbers 69, 70, 71, and 85 in table 2.1. Additional polluting activities of the company are included, along with other firms, under source numbers 11 and 51. For this particular study, the emissions from Granite City Steel were disaggregated from the other firms included under source numbers 11 and 51; thus, six separate sources of pollution were identified for the steel company, each characterized by the same diffusion factor m_j. Given the 1963 level of abatement, these six polluting activities would contribute an estimated total of 14 million pounds of particulates, 29 million pounds of sulfur dioxide, one million pounds of nitrogen oxides, and lesser quantities of carbon monoxide and hydrocarbons to the 1975 emission flow in the St. Louis airshed.

As noted above, the computed value of m_j for Granite City Steel Company, given its actual location (six miles northeast of the CAMP station), is 677×10^{-10} micrograms per cubic meter per pound. Table 4.6 contains alternative values of this diffusion constant for different locations of the company. Linear Programming Model V was run for each value of m_j (which represents six entries in the **M** matrix), and a least-cost solution was determined for each location. It can be seen from table 4.6 that if Granite City Steel Company were not located in the airshed ($k = \infty$), the total cost of abatement (which includes $431,000 for the base-year level of abatement by the company itself) would be $1.2 million less than the annual cost given the actual location (data corresponding to the actual location of the steel company are in boldface in table 4.6). If the company were located eight miles northeast of the CAMP station (two miles farther than its actual location), its annual control costs under the least-cost solution would be $.2 million less (than at six miles) and the total control costs for all sources $.5 million less. Alternatively, if the steel company were two miles closer to the CAMP station than it is now, its control costs in the least-cost solution would be $.3 million higher and the abatement costs for all sources together $1.1 million higher. The increases in abatement costs associated with the two-mile reduction in distance exceed the decreases in abatement costs for the two-mile increase in distance for two reasons: (1) the impact of a source on the concentration at the receptor increases geometrically as the distance of separation decreases—this is apparent in figure 4.3, in which the vertical axes begin at the concentrations attributable to all sources other than the steel company, given the 1963 level of abatement—and (2) the total cost of abatement increases at a constant or an increasing rate as the required reductions in concentration levels increase. Thus, when the distance of separation is reduced to only two miles northeast,[16] the total cost of abatement increases by $7.1 million. If the location were reduced to one mile northeast, the diffusion constant would triple to $15,466 \times 10^{-10}$; and the air quality goal for sulfur dioxide could not be met with the technology defined in table 2.1.

Table 4.6

Total annual cost of abatement in 1975 for various locations of Granite City
Steel Company

(1) Distance from the CAMP station	(2) Direction from the CAMP station	(3) Diffusion factor (micrograms per cubic meter per pound of pollutant)	(4) Total cost of abatement for the steel company (dollars)	(5) Total cost of abatement for all sources including the steel company
∞	NE	0	431,000	43,149,000
10 miles	NE	256×10^{-10}	559,000	43,680,000
8 miles	NE	392×10^{-10}	559,000	43,901,000
6 miles[a]	**NE**	$\mathbf{677 \times 10^{-10}}$	**754,000**	**44,352,000**
6 miles	SW	960×10^{-10}	754,000	44,724,000
6 miles	SE	1021×10^{-10}	754,000	44,825,000
6 miles	NW	1227×10^{-10}	1,059,000	45,135,000
4 miles	NE	1451×10^{-10}	1,059,000	45,420,000
2 miles	NE	5136×10^{-10}	1,385,000	52,527,000

[a]Data for the actual location of the steel company is shown in boldface (see text).

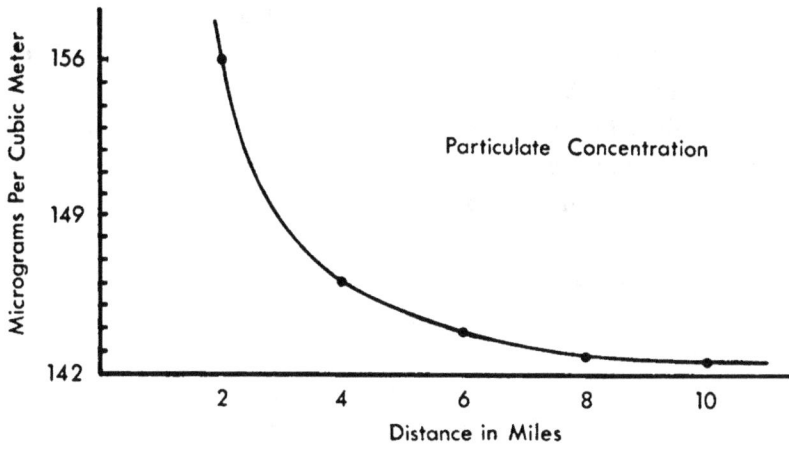

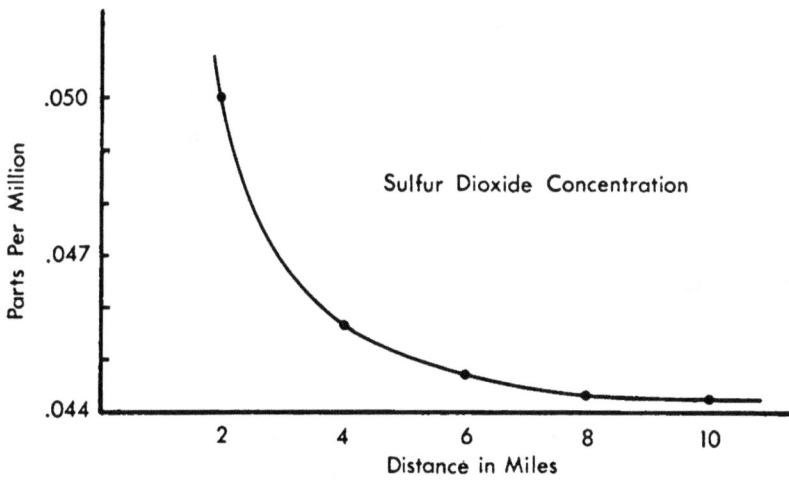

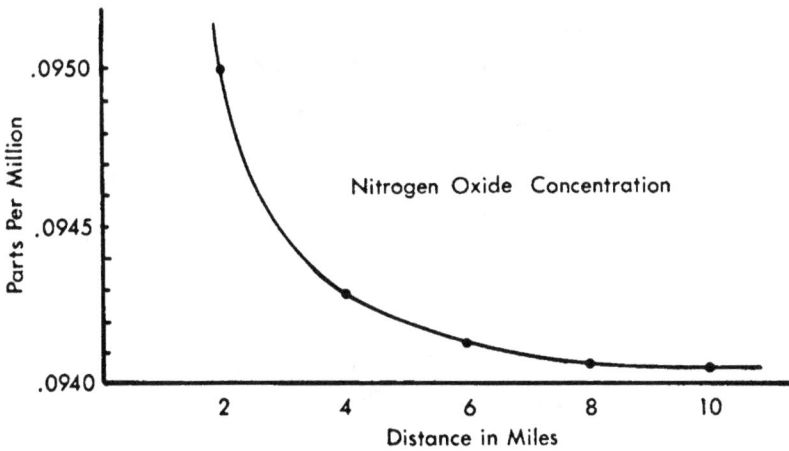

Figure 4.3
Pollutant concentration from Granite City Steel Company for various distances
northeast of the CAMP station

It is interesting that the location of the steel company northeast of the CAMP station is fortuitous. Had the plant been located at the same distance from the receptor, but in the southwest, southeast, or particularly in the northwest direction, its impact on the pollutant concentrations would have been greater (see table 4.6). This is the case because winds from these other directions are more frequent and/or their annual average velocities are less.

The data in table 4.6 suggest that the selective location of polluting sources can significantly reduce the total cost of emission abatement. However, the incremental transportation costs associated with relocation are ignored here.[17]

Guldmann has developed an empirical model (see table 4.7) in which *new* factories are located on vacant industrial sites in such a way as to minimize the total cost of emission control. While Guldmann avoids the costs of relocating existing factories, he still does not account for any incremental transportation or other costs that would be incurred if the location assignments based entirely on pollution control strategy were different from the locational decisions based on other cost minimizing considerations. In the Guldmann model it is the air quality in the individual residential zones that matters; this has advantages over Model V, in which the emphasis is on air quality at a single, fixed receptor point.

Shepard has also constructed an empirical model based on locational strategy (see table 4.7). Because the seven electric power plants in the St. Louis airshed are in different directions from the CAMP station, there is some potential for pollution control by shifting the generating load between the plants, depending on prevailing weather conditions (wind direction, etc.). The locational costs in this particular case are load-shifting costs, which are easier to compute than incremental transportation costs.

Achieved Air Quality under Alternative Planning Models

A comparison of the solutions (see table 4.5) of Model III and Model V indicates that whereas the same set of standards are achieved at the CAMP station, there is a lesser reduction in emission flows for four of the pollutants. This result, which was also obtained by Atkinson and Lewis (1974, pp. 247–248), implies that air quality will be the same at the CAMP station under the Larsen or the Diffusion formula, but is likely to be better elsewhere in the airshed using the Larsen formula. This is illustrated graphically by the hypothetical curves in figure 4.4.

Table 4.7

Empirical models of air quality management incorporating locational selectivity

Model	Shepard	Guldmann
Empirical application	St. Louis (1970)	Haifa, Israel (in some future year in which population has doubled)
Air quality indicator	One-hour concentration of sulfur dioxide	Annual average sulfur dioxide concentration in micrograms per cubic meter
Receptor points	Enough to generate pollution isopleths by a computer program	Sixteen (one in each residential zone)
Concentration/ emission relationship	Diffusion model	Diffusion model
Objective	Minimize exposure damage (100 persons suffer ten cents damage for one hour of exposure to one part per hundred million concentration) plus load-shifting costs	Minimize total fuel costs (transportation costs associated with strategic location of new plants are ignored)
Types of sources	Fourteen generating units in a total of seven electric power plants	Fifty-one polluting plants (representing eight two-digit industrial classifications)
Abatement alternatives	Shifting the generating load between power plants according to meteorological conditions	Three alternative fuels, differing in sulfur content, for each point source; and strategic assignment of new plants to appropriate locations, presently vacant, within fifteen industrial zones
Reference	Donald S. Shepard, "A Load Shifting Model for Air Pollution Control in the Electric Power Industry," *Journal of the Air Pollution Control Association*, 20, December, 1970, pp. 756–761.	Jean-Michel Guldmann, *Optimization Models for Air Pollution Control Strategies and Location of Industries: A Case Study of the Haifa Region*, Unpublished Master's Degree Thesis, Technion, Haifa, Israel, May, 1973.

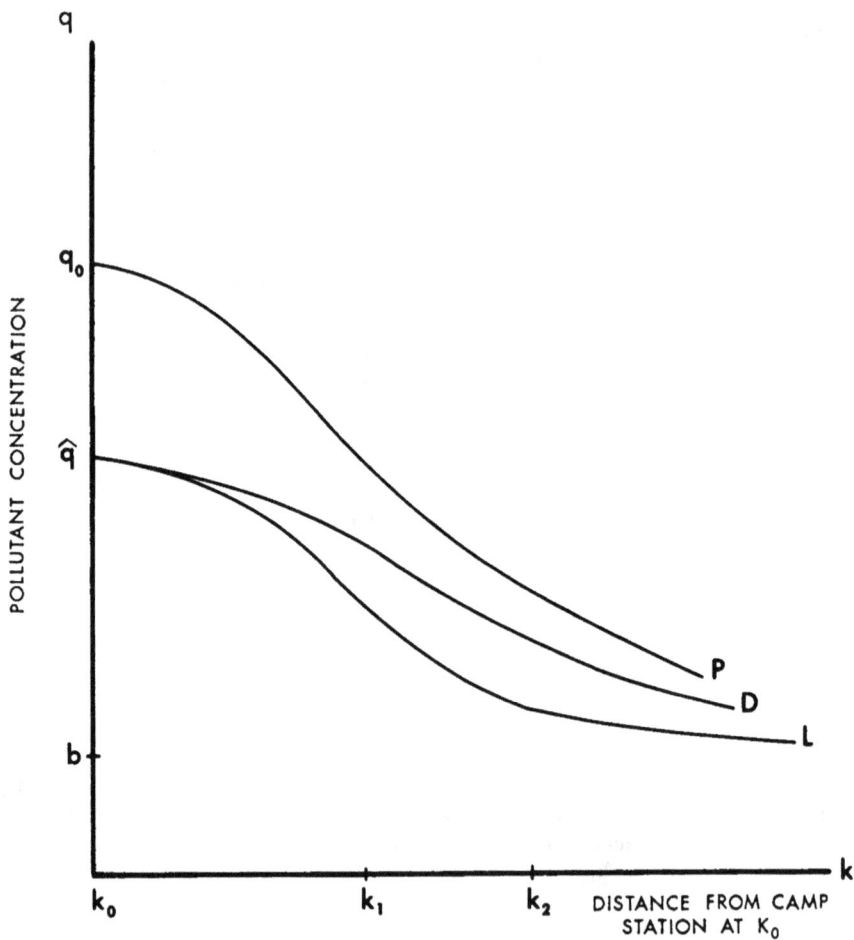

Figure 4.4
Hypothetical geographic pattern of pollutant concentrations before and after abatement

The horizontal axis in figure 4.4 represents distance from the CAMP station, which is located at $k = k_0 = 0$. Assuming that the CAMP station is located near the center of the airshed, where concentrations are relatively high, the level of concentration q decreases as k increases. The preregulatory pattern of concentrations is represented by the curve $q_0 p$ in figure 4.4. Letting $\hat{q}$ denote the air duality standard, the curve $\hat{q}L$ represents the pattern of air quality concentrations given efficient abatement based on the Larsen formula. The curve $\hat{q}D$ represents concentrations given efficient abatement based on the Diffusion formula. The former is likely (as is illustrated in table 4.5 for four of the pollutants) to result in greater reductions of total emissions and hence superior air quality away from the CAMP station. Note that all three curves become asymptotic to a horizontal line at b, which represents the background concentration.

Actually, the model based on the Diffusion formula could be constrained to a pattern of pollutant concentrations closer to curve $\hat{q}L$ in figure 4.4 by adding receptors at points such as k_1 and k_2 and setting the corresponding levels of q as supplementary standards. Furthermore, this would protect against the possibility, illustrated in figure 4.5, that the Larsen formula, which of necessity has a single standard for each pollutant, would yields a "hot spot" at k_1. This would not happen in a model based on the Diffusion formula and incorporating standards $\hat{q}$ at k_0 and $\hat{\hat{q}}$ at k_1. In fact, there is the problem of "hot spots" in the St. Louis airshed. At point L in figure 4.1 there is a valley containing some heavily polluting firms. Although these firms are complying with emission standards for St. Louis county, the concentrated volume of emissions and their confinement in this small valley are producing high pollutant concentrations. If the emission standards for these firms were based on a model with appropriate diffusion factors keyed to a local receptor, the problem, to the extent that it is a problem,[18] could be eliminated.

Relative Simplicity of the Larsen Formula

The Diffusion formula, with multiple receptor points, is a desirable planning tool. The data requirements, however, can be enormous. In effect, additional g, M, and q matrices are required for each receptor point. Although the Implementation Planning Program, the Atkinson and Lewis (see table 4.4), and the Shepard and Guldmann models (see table 4.7) all have numerous receptor points, only a subset of emission sources and one or two pollutants are included, so that these models remain tractable. Because the usefulness of

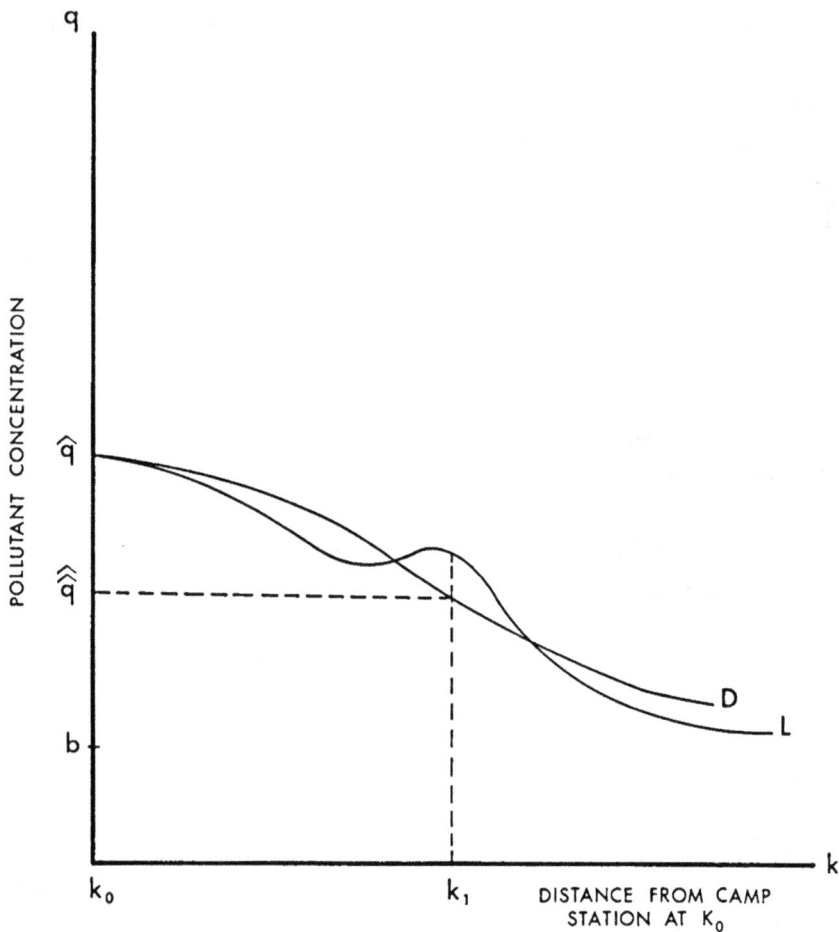

Figure 4.5
Hypothetical geographic pattern of pollutant concentrations with efficient abatement based on the Larsen formula as compared to a diffusion formula incorporating standards for two receptors

diffusion formulas depends on reliable estimates of meteorological variables, which are, in fact, stochastic, it is not always clear that the improvement in predictability is worth the increased computational effort.

Alternatively, a linear programming model based on the Larsen formula is relatively simple to implement. It is likely that such a model will indicate a greater reduction in total emissions than a model incorporating diffusion coefficients, and therefore, with the exception of some "hot spots," will probably result in better air quality throughout an airshed. Furthermore, there are periods of thermal inversion when an airshed *is* like a large box in which emissions accumulate and intermix; and a reduction in total emissions is crucial.

Unfortunately, the Larsen formula does not account for the fact that pollution sources may be dispersing over time (which implies a larger dimension for the Smith-Wanta box) or coming from taller stacks.[19] Consequently, the Larsen formula, too stringently applied, may straitjacket economic growth. This is illustrated by the data in table 4.8, which summarize the results of extending the Linear Programming Model of the St. Louis airshed to the year 1985.

The data for the 1985 model, Model VI, are the same as that in table 2.1 except that the source magnitudes have been projected to the later year. The same allowable flows are used, and as a result the required percentage reductions in emissions are greater for 1985 than for 1975. Whereas the cost of the base-year level of abatement increases by 4.4 percent per year (which is roughly equal to the overall annual increase in production), the cost of incremental abatement for meeting the allowable flows increases by 9.2 percent per year.[20] It may be, however, that pollution sources are dispersing over time and that the Larsen constants should be recalculated using data based on a later year than 1963. For the Larsen formula to be a valid one, it should be updated as frequently as is necessary.[21]

Chapter Summary

In this chapter we have examined various methods for relating emission flows to pollutant concentrations. The most simple approach is the Box formula, which implies a linear relationship between total emissions of a pollutant and the concentration of that pollutant. An alternative linear relationship, the Larsen formula, is estimated statistically; and if the geographic

Table 4.8

Percentage reductions in emissions required in 1975 and 1985 under the Larsen formula

(1) Pollutant	(2) Allowable emissions[a] (millions of pounds)	(3) Projected emissions in 1975 (millions of pounds)	(4) Percentage reductions in emissions required in 1975	(5) Projected emissions in 1985 (millions of pounds)	(6) Percentage reductions in emissions required in 1985
Carbon monoxide	2,335	4,202	44%	5,965	61%
Hydro-carbons	994	1,518	35%	2,188	55%
Nitrogen oxides	304	416	27%	596	49%
Sulfur dioxide	400	1,389	71%	1,945	79%
Particu-lates	136	300	55%	361	62%
Total annual cost of base-year level of abatement		$11.2 million		$17.3 million	
Total annual least cost of incremental abatement		$35.3 million		$85.3 million	

[a]See Table 4.2

Table 4.9

Total cost of abatement for versions of the Linear Programming Model referred to in chapter 4

Version	Special case	Total cost of abatement (millions of dollars)
Model III	Standard model based on the Larsen formula	46.5
Model IV	50 percent confidence of 99 percent probability of achieving the legal standards	71.3
Model IV	95 percent confidence of 99 percent probability of achieving the legal standards	88.0
Model V	Standard model based on the Calibrated Diffusion formula	44.4
Model V	Granite City Steel Company located northwest rather than northeast of the CAMP station	45.1
Model VI	Standard model based on the Larsen formula and 1985 source magnitudes	102.6

pattern of emission sources is stable, can be used to predict concentrations within a reasonable range of accuracy.

The total cost of abatement in the St. Louis airshed in 1975, based on the Linear Programming Model incorporating the Larsen formula, is $46.5 million. According to the Stochastic formula, however, the above total cost may be expected to achieve the required air quality standards with a confidence of only 50 percent that the probability of so doing is 50 percent. Higher degrees of confidence and probability require greater reductions of emissions and, hence, higher total costs of abatement. The total costs obtained with alternative versions of the Stochastic model (IV) are included in table 4.9.

The Diffusion formula accounts for the geographic location of pollution sources with respect to the receptor station. Although the total cost of abatement obtained with Model V, which incorporates the Diffusion formula, is less than the cost obtained with Model III, situations are conceivable in which the Diffusion formula would be associated with a higher total cost of abatement.

An economic model based on a Diffusion formula allows for pollution control strategy based on locational shifting of sources as well as technological abatement. This is demonstrated using a version of Model V in which alternative locations of a major source of pollution are simulated, thereby increasing or decreasing the total cost of abatement for all sources. If this major source had been located northwest of downtown St. Louis rather than northeast, the preregulatory concentrations at the CAMP station would have been higher and the total cost of abatement in 1975 would have been $700,000 more.

The Larsen formula is easlier to implement than the Diffusion formula and may be quite adequate for policy making. Furthermore, an abatement strategy based on the Larsen formula is likely to result in better air quality in outlying areas. However, as the results of Model VI demonstrate, the Larsen formula can become overly restrictive and, accordingly, should be revised as the geographic pattern of pollution sources changes over time.

5

OPTIMAL POLLUTANT
CONCENTRATIONS

The economic conditions for an optimal pollutant concentration can be expressed in terms of benefit-cost analysis. For reasons given in this chapter, the conventional representation of a continuous, downward sloping, marginal-benefit-of-abatement curve is valid only under special assumptions. If the air quality concentrations used as requirements in the Linear Programming Model are economically efficient, it follows that each pollutant shadow price must be equal to the marginal benefit of abating the corresponding pollutant. For purposes of verification, some outside estimates of pollutant damage are compared to the shadow prices derived from the Linear Programming Model.

A model for minimizing the sum of pollution damages plus abatement costs is described in this chapter. With this model, it is shown that the set of optimal pollutant concentrations changes with the rate of interest on capital invested in pollution control equipment.

The conditions for economic efficiency in the multipollutant case are simplified when air quality is expressed in terms of a single pollution index. In the case of a linear index system, an optimum is described by simultaneous divisible and edge conditions. Several pollution index systems are used in this chapter to determine benefit-effective pollutant concentrations for a predetermined abatement budget. In the case of sulfur dioxide and particulates, the benefit-effective concentrations for the St. Louis airshed are close to the legal standards. This analysis discloses a potential problem of nonconvexity in the relationship between the technology of abatement and curvilinear pollution index systems.

Benefit-Cost Analysis

In his pioneering work on the social costs of air pollution, Ridker (1967, pp. 4–6) suggested that the optimal standard for a pollutant is that concentration at which the marginal benefit of abatement equals the marginal cost of abatement. This is represented in figure 5.1 by the intersection at q^* of the marginal benefit (MBA) and marginal cost (MCA) curves. Our interpretation of benefits is based on equation (1.8) in chapter 1. For any level of air quality, the marginal benefits of abatement are $(-c_2)(U_q^1/U_2^1 + U_q^2/U_2^2)$.

This represents the dollar value that households place on the damage caused by the marginal unit of pollution.

The marginal benefit curve is somewhat analogous to the market demand curve for a private good. The price on the demand curve represents the marginal value of the private good to each household that purchases the good at that price. Whereas this price is the marginal value to each individual purchaser, the vertical axis in figure 5.1 measures the marginal value to all households combined.

There is a technical problem with the above analogy. The market demand curve assumes that all other prices are constant. Because abatement of air pollution occurs at the source, marginal costs of private goods, and hence selling prices, increase as emission rates are reduced. Accordingly, the value of the dollar, which is used to measure benefits, is not the same for all levels of air quality; and the marginal-benefit-of-abatement curve shifts as more costly pollution control methods are implemented. Thus the continuous benefit function in figure 5.1 is valid for a single set of abatement processes (see Kohn, 1975b, pp. 44–52, 71–73).

A market demand curve is a partial equilibrium construct in that a single price is allowed to vary. In the case of air pollution control, the marginal costs of private goods increase with the marginal cost of abatement, and a general equilibrium approach is essential. One of the advantages of the Pure Abatement model, as developed in chapter 1, is that general equilibrium can be characterized by some of the simplifying assumptions of partial equilibrium analysis. Thus, for example, we can represent the marginal benefits of abatement as a continuous curve in which benefits are measured in units of the composite good. This is illustrated in figure 5.2.

In the Pure Abatement model, the technology is such that the relative prices of final goods do not change. The marginal-benefit-of-abatement curve, which can be expressed in units of the composite good, is a summation of the corresponding curve for each of the households. This vertical summation for a three-household economy is illustrated in figure 5.2: the diagram is analogous to figure 1.3 in chapter 1, except that the vertical axis represents units of the composite good per unit of pollutant concentration. The horizontal segments of the marginal cost of abatement function in figure 5.2 correspond to the consecutive slopes of the frontier abcd in figure 1.3 and are measured in units of the composite good.

The representation in figure 5.2 is valid regardless of whether pollution

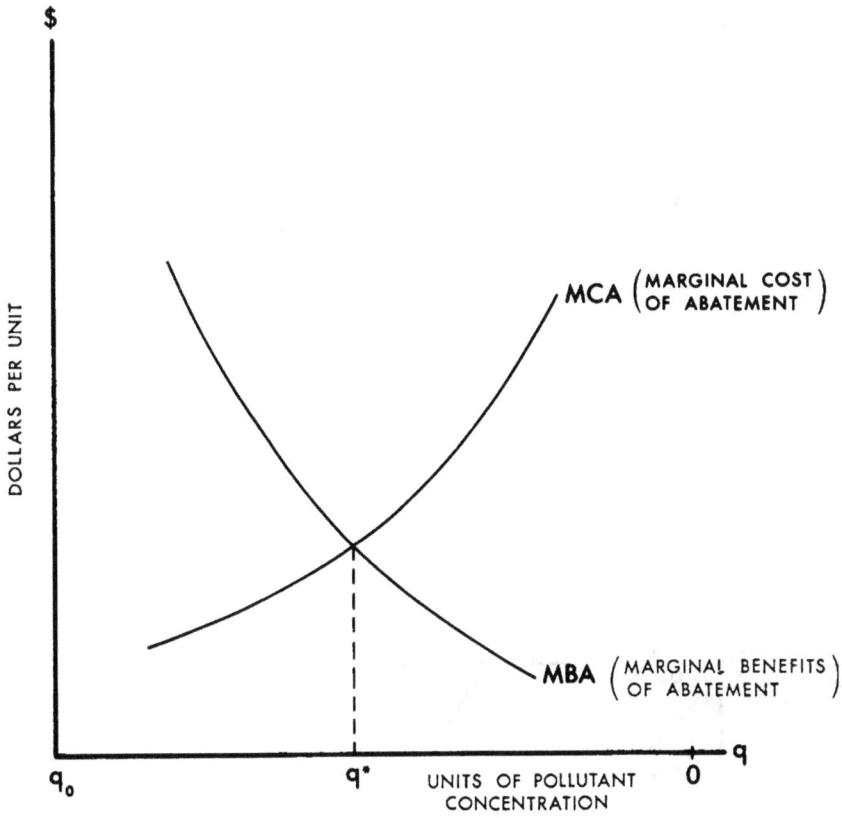

Figure 5.1
The optimal concentration of a pollutant

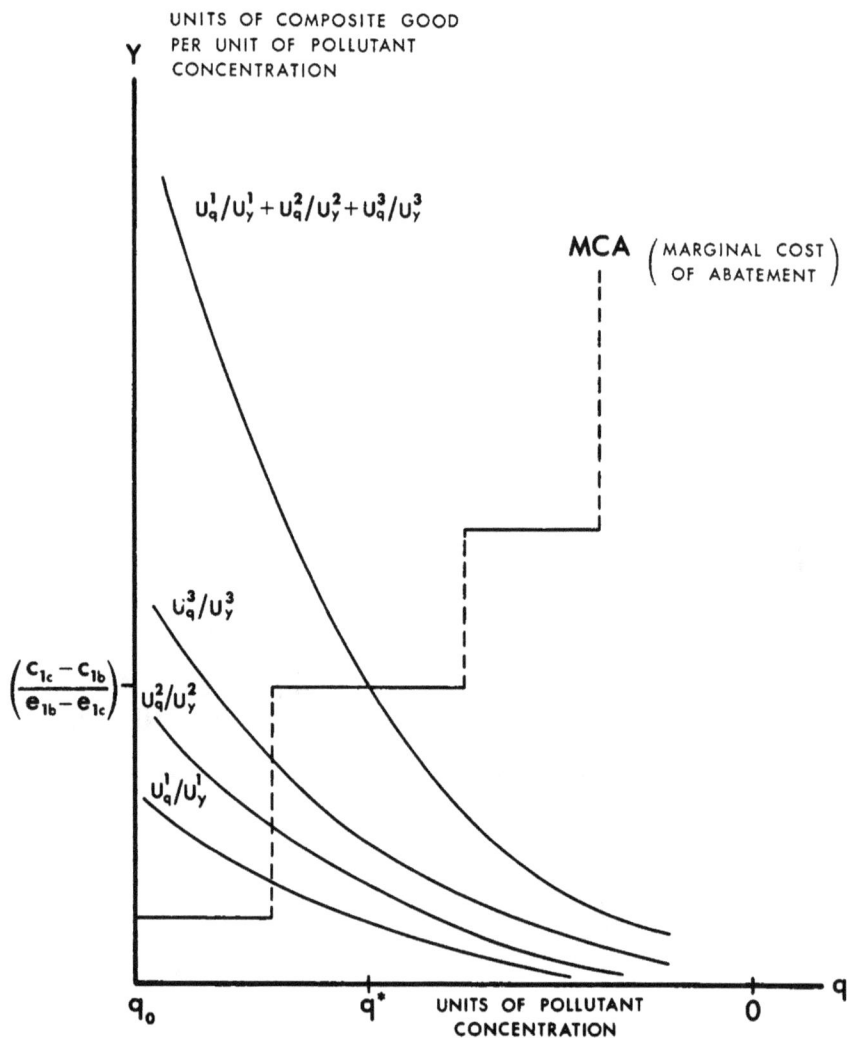

Figure 5.2
Summation of household marginal rates of substitution of the composite good for air pollution

control is achieved through emission standards or emission fees. However, the distribution of income would probably be different under the two programs; and this would result in different marginal-benefit-of abatement-curves. Provided that the indifference curves have conventional properties and the rule for distributing fee revenue to households is independent of the abatement technology, the marginal benefit curves may be assumed to be continuous and downward sloping, as illustrated.

If there are several pollutants and there are interdependencies between them in abatement, then figure 5.2 will be invalid. This kind of interdependency is illustrated in table 5.1. The data in this table were obtained by running Model I with a sequence of allowable flows on particulates alone. As the allowable flow of particulates decreases (or the required abatement increases) there is joint abatement of other pollutants (see columns 3, 4, 5, and 6). The benefits of this joint abatement would not be included in the marginal benefit curves in figure 5.2.

Furthermore, the shadow prices in column 2 overstate the costs of abatement. When requirements are imposed on the remaining pollutants, some of the reduction in particulates is a by-product of the abatement of the other pollutants, while part of the cost of particulate control is allocated to these other pollutants. Accordingly, the particulate shadow prices are lower in column 7 of table 5.1 than in column 2. Thus, a benefit-cost analysis based on the One-Pollutant model, as illustrated in figure 5.2, would indicate less abatement of particulates as optimal than would the Multipollutant model.

Columns 8, 9, 10, and 11 of table 5.1 show the shadow prices for the other pollutants in the Multipollutant model. (The shadow prices from the orginal solution of Model I are in boldface type.) Note that as the particulate requirement is increased, the shadow prices for hydrocarbons and nitrogen oxides decrease. This characteristic of the Multipollutant model was observed in the succession of shadow prices in the numerical example in the appendix of chapter 1.

We have demonstrated that the marginal benefits of abatement cannot be measured in dollars and then represented by a continuous curve, as in figure 5.1. In the Pure Abatement model, a continuous curve is valid provided that benefits are meaured in units of the composite good and there is no joint abatement of other pollutants. This does not mean that we cannot use dollar values to define the conditions of economic efficiency for an

Table 5.1

The One-Pollutant and the Multipollutant models compared for various levels of particulate abatement

Particulate abatement (millions of pounds) (1)	The One-Pollutant model				
	Particulate shadow price (cents) (2)	Joint-abatement (millions of pounds)[b]			
		Carbon monoxide (3)	Hydro-carbons (4)	Nitrogen oxides (5)	Sulfur dioxide (6)
220	63.249	115	291	12	258
210	28.620	112	283	7	162
200	23.745	101	245	7	161
190	23.745	82	186	7	161
180	20.000	68	137	5	100
170	12.277	63	137	3	51
163.804	9.184	62	137	3	51
160	9.184	62	137	3	51
150	8.598	60	129	3	51
140	6.160	58	128	3	47
130	6.160	48	128	1	23
120	4.800	39	127	0	1
110	3.906	39	127	0	1
100	3.029	39	127	0	1
90	3.029	39	127	0	1
80	2.969	39	127	0	1
70	1.979	37	121	0	1
60	1.979	19	61	0	0
50	1.979	1	2	0	0
40	1.913	0	0	0	0
30	1.696	0	0	0	0
20	1.537	0	0	0	0
10	.133	0	0	0	0

[a]The abatement requirements in the Multipollutant model are as follows: carbon monoxide, 186.701 million pounds; hydrocarbons, 524.265 million pounds; nitrogen oxides, 111.945 million pounds; and sulfur dioxide, 989.163 million pounds.

[b]There are several abatement activities that show a net profit. The emission abatement associated with these activities is not included in columns (3) through (6).

The Multipollutant model[a] (shadow prices in cents)

Particulates (7)	Carbon monoxide (8)	Hydro- Carbons (9)	Nitrogen oxides (10)	Sulfur dioxide (11)
28.673	.933	0	28.362	2.193
21.274	.920	.063	29.411	2.193
21.274	.920	.063	29.411	2.193
12.277	.593	1.668	31.558	2.193
9.184	.480	2.220	32.297	2.193
8.598	.459	2.324	32.436	2.193
7.748	**.428**	**2.476**	**32.639**	**2.193**
7.748	.428	2.476	32.639	2.193
3.906	.428	2.476	33.165	2.193
2.969	.428	2.476	33.294	2.193
2.969	.428	2.476	33.294	2.193
1.913	.428	2.476	33.438	2.193
1.698	.428	2.476	33.468	2.193
1.389	.428	2.476	33.964	2.193
.424	.428	2.476	37.054	2.193
0	.428	2.476	37.899	2.193

equilibrium allocation of inputs and outputs. Given an equilibrium allocation, the vector of pollutant concentrations is optimal if the dollar benefit of reducing any of the concentrations by a small amount is less than the corresponding shadow price, or if the incremental pollution damage associated with a small increase in any of the concentrations exceeds the shadow price. This multiple equilibrium is illustrated in figure 5.3 using data from Model I and relating marginal benefits of abatement and shadow prices to pounds of emissions.

Pollutant Shadow Prices as Estimates of Marginal Damage

If the pollutant requirements in the Linear Programming Model are optimal and its parameters are correct, the shadow prices of the concentrations should equal the marginal benefits of abatement of the respective pollutants. According to Kneese and Bower (1968, pp. 133–134), each pollutant shadow price should give "an indication of what a small change in the standard and the accompanying physical effects must at least be worth."

The pollutant shadow prices from Model III and Model V are contained in columns 3 and 6 of table 5.2.[1] For uniformity, the shadow prices of the gaseous pollutants are converted from parts per million to micrograms per cubic meter (see note 2 in chapter 4); the uniform shadow prices corresponding to column 3 are listed in column 4 and those corresponding to column 6 in column 7.

There have been several pollution control models based on estimates of per capita damage per unit of pollutant concentration. Several of these studies are included in table 5.3. Wilson and Minnotte combine benefit and cost data to determine optimal concentrations for the Washington, D.C., area. The damage data used in the Chicago Project (which are derived from recent studies by Cohen et al. (1974, pp. 61–86) and Zerbe and Croke (1975)) are used to evaluate the economic efficiency of pollution control regulations in the Chicago area. It is interesting to compare the shadow prices in table 5.2 with the damage figures used in the above studies. To do so, the shadow prices in columns 4 and 7 are each divided by the population in the St. Louis area, which was approximately 2,500,000 in 1975. (Note that this population figure is somewhat less than the earlier estimate for 1975, which is the magnitude for source 91 in table 2.1.) The resulting per capita damage per microgram figures are contained in columns 5 and 8.

The damage estimates used in the Chicago Project (1975, p. 24) are

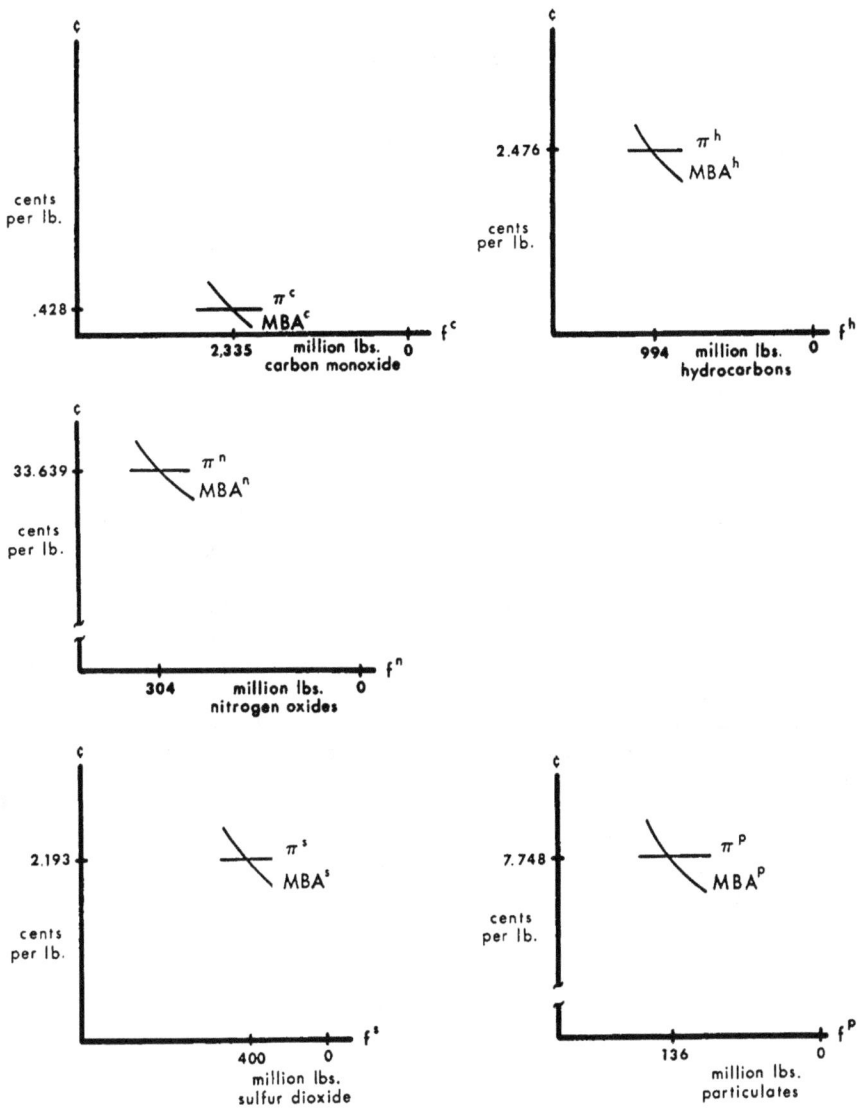

Figure 5.3
Economic efficiency in the five-pollutant case

Table 5.2
Pollutant shadow prices and the benefits of abatement (dollars)

(1)	(2)	(3)	(4)
			Pollutant shadow prices from Model III per microgram per cubic meter[b]
		Pollutant shadow prices from Model III in given units of concentration[a]	
Pollutant	Pollutant standards		
Carbon monoxide	5.0 ppm	2,000,000	1,700
Hydrocarbons	3.1 ppm	15,388,000	22,700
Nitrogen oxides	.069 ppm	1,435,800,000	900,200
Sulfur dioxide	.02 ppm	439,100,000	162,000
Particulates	75 μg/m	239,100	239,100

[a]These shadow prices relate to parts per million for gaseous pollutants and micrograms per cubic meter for particulates. The shadow prices in column 3 are based on a model that incorporates the Larsen formula. Multiplication by the Larsen constant m^i (see table 4.2) gives the pollutant shadow prices in table 3.3. The shadow prices in column 6 are based on the model that incorporates the Diffusion formula.
[b]The shadow prices for the gaseous pollutants are converted from ppm to μg/m^3 by multiplying by the conversion factors in note 2 of chapter 4.
[c]Based on a 1975 population of 2,500,000 people in the St. Louis airshed.
[d]See table 5.3. These estimates denote average damage per person per microgram of pollutant per cubic meter of air. No damage figures were given for carbon monoxide and hydrocarbons.
[e]"n" stands for less than one-half cent.

(5)	(6)	(7)	(8)	(9)
Implied marginal pollutant damage per capita[c]	Pollutant shadow prices from model V in given units of concentration[a]	Pollutant shadow prices from Model V per microgram per cubic meter[b]	Implied marginal damage per capita[c]	Average damage per capita estimates used in the Chicago Project[d]
n[e]	5,100,000	4,300	n	—
.01	0	0	0	
.36	174,300,000	109,300	.04	.80
.06	788,700,000	291,000	.12	1.00
.10	593,200	593,200	.24	2.00

contained in column 9 of table 5.2. These estimates are higher than those in columns 5 and 8, and there are two reasons why this should be the case. The shadow prices corresponding to optimal standards would equal marginal damage *after* abatement, and assuming that marginal damage declines as air quality improves, we would expect marginal damage after abatement to be less than average damage.[2] Second, the estimates of damage in the Chicago Project relate to actual exposure, whereas the St. Louis models relate damages throughout the airshed to concentrations at the CAMP station. In fact, the bulk of the population in St. Louis is exposed to much lower concentrations than those at the CAMP station.[3] It may be that these two factors provide a satisfactory explanation for the difference between the damage estimates in columns 5 and 7 and those in column 9. However, the large contrast may also indicate that certain unit costs of abatement in the Linear Programming Model are too low. Alternatively, it may be the case that the pollutant standards for the St. Louis airshed are not sufficiently stringent.

The damage estimates in table 5.2 are based on pollutant concentrations. If the Larsen formula is appropriate for planning, it follows that there is some total annual emission flow that that is optimal for each pollutant. Accordingly, the shadow prices corresponding to these optimal flows would

Table 5.3
Empirical Benefit-Cost models

Model	Wilson and Minnotte	Chicago project
Empirical application	Washington, D.C. (1968)	Chicago, Illinois (1980)
Air quality indicator	Annual average particulate concentration in micrograms per cubic meter	Annual average concentration in micrograms per cubic meter for particulates, sulfur dioxide, and nitrogen oxides
Receptor points	32	Many
Concentration/ emission relationship	Diffusion model	Diffusion model
Objective	Maximize total benefits of abatement minus total costs. The marginal benefits of particulate abatement are an estimated $1.85 per person per microgram per cubic meter.	Maximize total benefits (estimated at the following annual per person values for each microgram per cubic meter reduction in exposure: $2.00 for particulates, $1.00 for sulfur dioxide, and $.80 for nitrogen oxides) minus total costs of abatements for current regulations in the Chicago area and for alternative regulations.
Types of sources	131 point (mostly large coal burning furnaces) sources	Power plants, industries, and residential furnaces
Abatement alternatives	Lower ash coal, scrubbers, precipitators, and substitute fuels	Lower sulfur coal, flue gas desulfurization, coal gasification, increasing stack height, scrubber, afterburner, lower sulfur oil, etc.
Reference	Richard D. Wilson and David W. Minnotte, "A Cost-Benefit Approach to Air Pollution Control," *Journal of the Air Pollution Control Association*, 19, May, 1969, pp. 303–308.	*Environmental Pollutants and the Urban Economy, Phase I*, University of Chicago, Center for Urban Studies, and Argonne National Laboratory, 1976.

equal the marginal damages per pound of emissions. Some work has been done on estimating damages per pound of particulates. Using damage figures from Waddell (1973), Atkinson and Lewis (1976, p. 371) estimate that attainment of the 75 μg/m^3 particulate standard in the St. Louis airshed would yield average benefits of $251 per ton of particulate abatement. This estimate, which is equivalent to 12.5¢ per pound, compares favorably to the particulate shadow price of 7.7¢ per pound in Model I (see table 3.3). The above estimate by Atkinson and Lewis is their medium figure. They also provide a low estimate equivalent to 6.3¢ per pound and a high estimate equivalent to 18.9¢ per pound. Thus, the interpretation of the particulate shadow price from Model I as a measure of pollutant damage appears to be consistent with the damage estimates of Atkinson and Lewis.

A Benefit-Cost Version of the Linear Programming Model

If the concentrations $\hat{q}$ in Model III are optimal, the corresponding pollutant shadow prices π_{III}, would equal the marginal benefits of abatement. If we assume that marginal benefits are constant, it follows that the total benefits in 1975 of reducing concentrations in the St. Louis airshed, from those predicted with the Larsen formula (see table 4.5) to the optimal standards, would be

$$
\begin{aligned}
(9.0 - 5.0)\ (\$2,000,000) + (3.9 - 3.1)\ (\$15,388,000) \\
+ (.094 - .069)\ (\$1,435,800,000) \\
+ (.069 - .020)\ (\$439,100,000) + (128 - 75)\ (\$239,100) \\
= \$90,393,600.
\end{aligned}
\tag{5.1}
$$

This estimate of benefits is equivalent to approximately $36 per capita, as compared to a national figure of $50 (based on data in Leung and Klein, 1976, p. 28). If pollutant damages increase with concentrations, however, as in figure 5.1, the estimate of annual benefits would be higher than the $90.4 million in (5.1). This estimate of total benefits for the St. Louis airshed substantially exceeds the incremental total cost of abatement of $35.3 million (see table 3.1).

Ridker (1967, pp. 4–6) has shown that the Benefit-Cost model can be viewed as a Damage-Cost model, in which the optimal pollutant concentration is that concentration at which the sum of total pollution damages plus total abatement costs is minimized. Assuming that π_{III}, the 5 × 1 vector of pollutant shadow prices from the solution of Model III, represents both marginal and average damage for the respective pollutants, the Ridker version of the Linear Programming Model is

Minimize $(\pi_{III})q + Cx$
subject to $q - mex = b$, $\qquad\qquad\qquad\qquad\qquad$ (5.2)
$\qquad\qquad ux = s$
$\qquad\qquad q, x \geqslant 0$.

This version of the model, called Model VII, was used to test the relationship of the variable q to the interest rate that represents the opportunity cost of capital (see cost calculations in (2.10) above). Because pollution abatement is capital intensive, it could be expected that the higher the opportunity cost of capital, the greater (that is, the less stringent) be the would optimal pollutant concentrations.

The results of Model VII are contained in table 5.4. Observe that the optimal standards in column 4 differ from the given standards in column 2. This is the case because the solution of Model VII is an *integer* solution without divisibilities. This occurs because the shadow prices from the computer print-out for Model III are rounded out and differ from the exact shadow prices. The fact that all of the solutions of Model VII are corner solutions will be discussed later in this chapter.

Model VII was used to test the effect of a small change in the interest rate on the choice of air quality standards.[4] When the opportunity cost of capital was reduced from 10 to $7\frac{1}{2}$ percent, more stringent standards for three of the pollutants were indicated. However, the interdependencies in abatement between pollutants were such that the relative cost of controlling sulfur dioxide actually increased and the optimal standard at the $7\frac{1}{2}$ percent rate was above that at the 10 percent rate.[5] When the opportunity cost of capital was increased from 10 to $12\frac{1}{2}$ percent, the abatement budget declined and a less stringent set of standards was optimal.

The major assumptions in this analysis are that the dollar damage of one unit of a particular pollutant is constant, at least over the relevant range of concentrations, that it is independent of changes in the level of other pollutants, and that the income effect (resulting from the change in the total cost of abatement) on the valuation of damages is insignificant.

In Model VII both q and z are variables. When the opportunity cost of capital decreases, it is efficient to have more stringent air quality standards and to spend more money on abatement. When the opportunity cost of

Table 5.4

Optimal concentrations from Benefit-Cost Model VII in which the benefits of abatement are equal to the pollutant shadow prices from Model III

(1) Pollutant	(2) Given standards in Model III	(3) Optimal standards for a 7½% opportunity cost of capital	(4) Optimal standards for a 10% opportunity cost of capital	(5) Optimal standards for a 12½% opportunity cost of capital
Carbon monoxide	5.0 ppm	4.6 ppm	4.7 ppm	7.3 ppm
Hydrocarbons	3.1 ppm	2.8 ppm	3.1 ppm	3.5 ppm
Nitrogen oxides	.069 ppm	.069 ppm	.069 ppm	.069 ppm
Sulfur dioxide	.020 ppm	.020 ppm	.019 ppm	.022 ppm
Particulates	75.0 $\mu g/m^3$	67.8 $\mu g/m^3$	76.8 $\mu g/m^3$	77.7 $\mu g/m^3$
Total annual cost of abatement[b]	$46.5[a]	$48.4	$47.2	$37.9
Invested capital	$233.5	$246.1	$236.0	$172.9

[a]A 10% opportunity cost of capital is used.
[b]Millions of dollars.

capital increases, the opposite is true. This is somewhat analogous to the case of a private good for which the price elasticity of demand is elastic (that is, greater than unity); as price decreases (increases), the quantity of money spent on the good increases (decreases). This result is consistent with observations (see Dorfman, 1977) that the demand for environmental quality is income elastic.

Pollution Index Systems

In chapter 1 we derived the mathematical conditions for an optimal combination of a composite good and the concentrations of two pollutants. The concentration of a given pollutant, however, will vary spatially within an airshed. Letting q^{ik} represent the concentration of the ith pollutant at the kth location, the utility functions of households would, in theory, include each of the perceived concentrations. The marginal rates of substitution for households would then include trade-offs of one pollutant concentration at one location for another concentration of the same or of a different pollutant at another location. The conditions for economic efficiency would include summations of individual household trade-offs and rates of transformation in production, as in equations (1.22) and (1.23) of chapter 1. At equilibrium the marginal cost of abatement for each q^{ik} would be equal to the marginal benefit of abatement.[6]

The conditions for economic efficiency, with so many dimensions of air quality, are complex. It is doubtful whether households would indeed be able to evaluate marginal changes in the concentration of each one of a set of pollutants, even at a single location. The conceptual as well as the practical complexity of the multipollutant case can be simplified by expressing the level of air pollution as a function of the individual pollutant concentrations. These could be the annual average concentrations as measured at the CAMP station or an average of concentrations measured at a number of stations throughout the airshed. There are, in fact, a number of such functions, generally called Pollution Index Systems, currently employed in the United States and Canada (see Ott and Thom, 1976). They range from nonlinear functions containing exponents to separable linear functions, and are being used by regulatory agencies to express the level of air quality as a single index number.

Let us consider a linear pollution index in which the annual average concentration of each pollutant is weighted by an estimate of the relative toxicity of that pollutant. This results in a function of the form

$$Q = w^1 q^1 + w^2 q^2 + w^3 q^3 + \cdots, \tag{5.3}$$

where Q is an overall measure of the severity of air pollution and the w^i are relative toxicity weights. Assuming that a function such as (5.3), based on annual average concentrations at some central receptor such as the CAMP station, has a numerical value that each household perceives as the overall level of air pollution, then the public demand for an improvement in air quality can be represented by a single summation of trade-offs.

To illustrate, consider our simple economy with two households, in which pollution originates during the production of an intermediate good used by all firms, in proportion α to their labor input. In this special case the vector of private goods can be represented by a single composite good called Y. Emission rates are related to concentrations as follows:

$$\begin{aligned}
q^1 &= e^1_{1a} x_{1a} + e^1_{1b} x_{1b} + e^1_{1c} x_{1c} + e^1_{1d} x_{1d} + b^1, \\
q^2 &= e^2_{1a} x_{1a} + e^2_{1b} x_{1b} + e^2_{1c} x_{1c} + e^2_{1d} x_{1d} + b^2.
\end{aligned} \tag{5.4}$$

The level of pollution that both households experience is

$$Q = w^1 q^1 + w^2 q^2, \tag{5.5}$$

where w^1 and w^2 are positive constants. The conditions for an economically efficient combination of Y_1, Y_2, and Q are derived from the Lagrangian expression,

$$\begin{aligned}
\mathscr{L} = \; & U^1(Y_1, Q) + \lambda_u[U^2(Y_2, Q) - \bar{U}^2] \\
& + \lambda_r[R - c_{1a} x_{1a} - c_{1b} x_{1b} - c_{1c} x_{1c} - c_{1d} x_{1d} - Y_1 - Y_2] \\
& + \lambda_1[\alpha(Y_1 + Y_2) - (1 - \alpha c_{1a}) x_{1a} - (1 - \alpha c_{1b}) x_{1b} \\
& - (1 - \alpha c_{1c}) x_{1c} - (1 - \alpha c_{1d}) x_{1d}].
\end{aligned} \tag{5.6}$$

Because the level of pollution is represented by a single number, no more than two processes would be used for making the intermediate good. Consequently, the optimal solution would include both divisible and corner conditions. Consider, for example, the case in which processes 1b and 1c are nonzero. The Kuhn-Tucker contitions for (5.6) yield the following equality:

$$-(U^1_Q/U^1_Y + U^2_Q/U^2_Y) = \frac{c_{1c} - c_{1b}}{w^1(e^1_{1b} - e^1_{1c}) + w_2(e^2_{1b} - e^2_{1c})}. \tag{5.7}$$

This states that the sum of the quantities of the composite good that each household in the economy would forego to reduce the pollution index by one unit should be equal to the rate of transformation in production between Q and Y.[7]

Condition (5.7) is illustrated in figure 5.4. In this diagram it is assumed that there are no background concentrations. The production-possibility frontier is abcd. The slopes at Q^* of the indifference curves $\bar{U}^1$ and $\bar{U}^2$ sum to the slope of the facet bc. This slope is the right-hand-side value in (5.7) and may be calculated from the coordinates of vertices b and c. (Those for vertex b are indicated in figure 5.4.)

The optimal index level Q^* can be achieved with alternative values of q^1 and q^2. Accordingly, there are efficiency conditions relating the ratio of pollutant toxicities to the rate of transformation between the corresponding pollutant concentrations. The convexity of the production-possiblity frontier of pollutant concentrations, given any fixed output of private goods, implies that in moving from one facet of the frontier to the next, thereby achieving lower concentrations of, say, pollutant-one in exchange for higher concentrations of pollutant-two, the absolute value of the shadow price of pollutant-one will increase while that of pollutant-two will decrease. We shall assume that the technology underlying (5.6) is such that

$$
\begin{aligned}
\pi^1_{abc} &> \pi^1_{bcd}, \\
\pi^2_{abc} &< \pi^2_{bcd},
\end{aligned}
\tag{5.8}
$$

where π^i_{abc} denotes the shadow price of the ith pollutant and the subscript indicates the three processes 1a, 1b, and 1c that generate the facet corresponding to that shadow price. Because the shadow prices are negative, it follows that the marginal rate of transformation of pollutant-two for pollutant-one increases from facet abc to facet bcd:

$$
\frac{\pi^1_{abc}}{\pi^2_{abc}} < \frac{\pi^1_{bcd}}{\pi^2_{bcd}}.
\tag{5.9}
$$

The Kuhn-Tucker conditions for an optimal solution of (5.6) yield the following inequality:

$$
\begin{aligned}
&\frac{(c_{1c} - c_{1b})(e^2_{1a} - e^2_{1b}) - (c_{1b} - c_{1a})(e^2_{1b} - e^2_{1c})}{-(c_{1c} - c_{1b})(e^1_{1a} - e^1_{1b}) + (c_{1b} - c_{1a})(e^1_{1b} - e^1_{1c})} \\
&\leq \frac{w^1}{w^2} \leq \frac{(c_{1d} - c_{1c})(e^2_{1b} - e^2_{1c}) - (c_{1c} - c_{1b})(e^2_{1c} - e^2_{1d})}{-(c_{1d} - c_{1c})(e^1_{1b} - e^1_{1c}) + (c_{1c} - c_{1b})(e^1_{1c} - e^1_{1d})}.
\end{aligned}
\tag{5.10}
$$

This is equivalent to

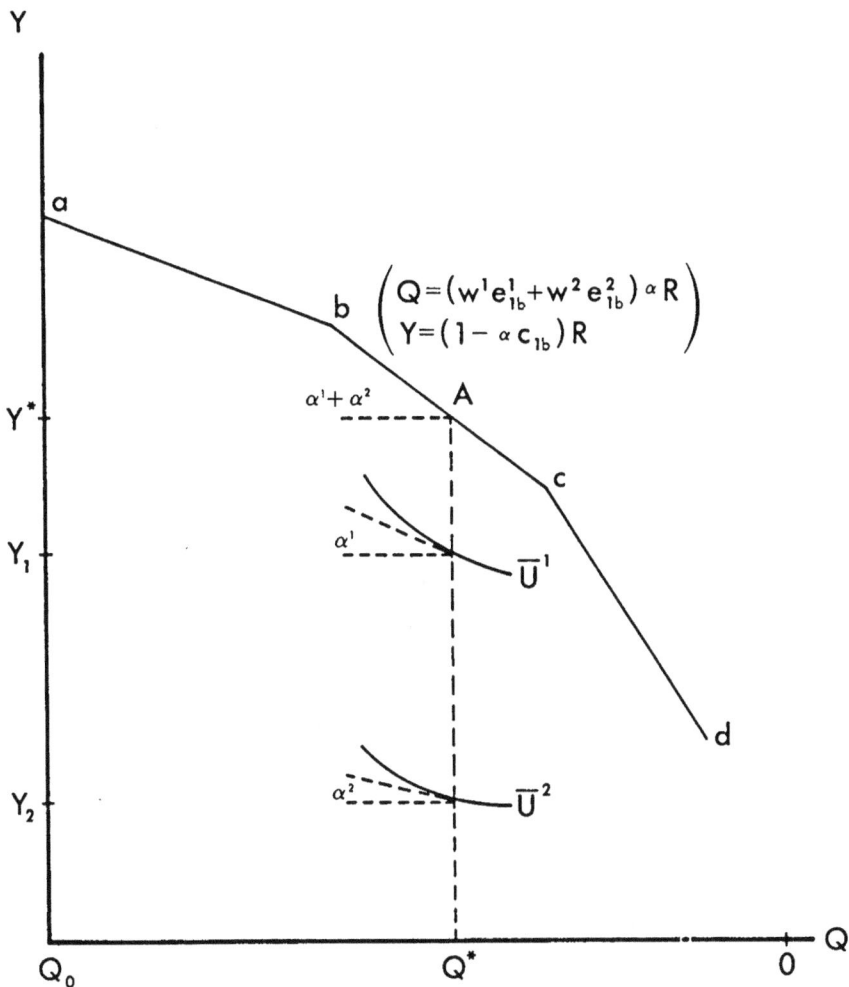

Figure 5.4
A Pareto optimal level of the pollution index and the composite good

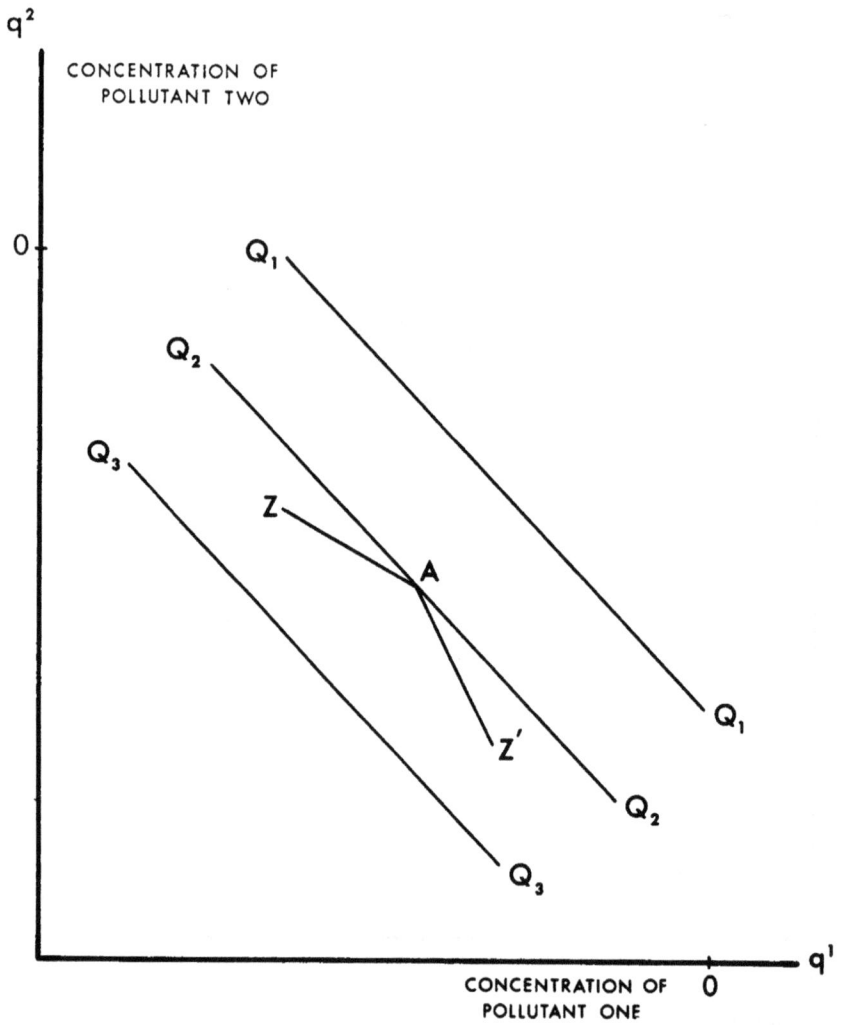

Figure 5.5
An optimal combination of two pollutant concentrations in an edge solution

$$\frac{\pi^1_{abc}}{\pi^2_{abc}} \leqslant \frac{w^1}{w^2} \leqslant \frac{\pi^1_{bcd}}{\pi^2_{bcd}}. \qquad (5.11)$$

This condition for an optimal combination of q^1 and q^2 states that the relative toxicity of the concentration of pollutant-one to that of pollutant-two is greater than the rate of transformation between them using processes 1a, 1b, and 1c, but is less than the rate of transformation using processes 1b, 1c, and 1d. It follows that (5.10) defines an edge solution.

Condition (5.10) is illustrated in figure 5.5. The frontier ZAZ′ denotes the available budget for abatement, given the optimal level of output Y^*. The line segment ZA denotes combinations of q^1 and q^2 obtainable with processes 1a, 1b, and 1c, while AZ′ denotes combinations attainable with processes 1b, 1c, and 1d. (Note that each facet represents a different basis matrix.) The straight line isoquants Q_3Q_3, Q_2Q_2, and Q_1Q_1 denote successively lower and therefore more desirable levels of the pollution index. The optimal value of Q is the value of the index isoquant tangent to ZAZ′. In figure 5.5 the point of tangency is A, and Q_2 is equivalent to Q^* in figure 5.3.

Benefit-Effectiveness

If an optimal abatement budget Z^* were predetermined, the economically efficient combination of pollutant concentrations $\{q^1, q^2\}$ would be the solution to the following linear programming problem:

Minimize $Q = w^1q^1 + w^2q^2$

subject to

$$q^1 - e^1_{1a}x_{1a} - e^1_{1b}x_{1b} - e^1_{1c}x_{1c} - e^1_{1d}x_{1d} = 0,$$
$$q^2 - e^2_{1a}x_{1a} - e^2_{1b}x_{1b} - e^2_{1c}x_{1c} - e^2_{1d}x_{1d} = 0, \qquad (5.12)$$
$$C_{1a}x_{1a} - C_{1b}x_{1b} - C_{1c}x_{1c} - C_{1d}x_{1d} \leqslant Z^*,$$
$$x_{1a} + x_{1b} + x_{1c} + x_{1d} = \alpha R,$$
$$x_{1a}, x_{1b}, x_{1c}, x_{1d} \geqslant 0.$$

The solution of (5.12) is graphically shown in figure 5.5. This same model can be used to determine the lowest possible pollution index value Q for *any* given abatement budget Z. As such, (5.12) is equivalent to a benefit-effectiveness model in which total benefits of abatement are maximized for a given total cost. Thus, benefit-effectiveness is the analogue of cost-effectiveness as typified in Model III. In that model the cost of abatement is minimized for a given set of concentrations.

The Linear Programming Model was transformed into a benefit-effec-

Optimal Pollutant Concentrations

tiveness version, called Model VIII. This model utilizes a linear pollution index that is minimized. In matrix notation, the model is

Minimize $\mathbf{Q} = \mathbf{wq}$
subject to
$$\begin{aligned}
\mathbf{q} - \mathbf{mex} &= \mathbf{b}, \\
\mathbf{ux} &= \mathbf{s}, \\
\mathbf{Cx} &= \mathbf{Z}^*, \\
\mathbf{q}, \mathbf{x} &\geqslant 0.
\end{aligned} \tag{5.13}$$

Model VIII was implemented with toxicity weights adapted from a study by Sterling, Pollack, and Phair (1967). In that study, pollutant concentrations in Los Angeles were correlated with length of hospital stay of some 30,000 area patients afflicted with allergic disorders and with heart, vascular, and respiratory infections and diseases. The hypothesis was that air pollution lengthens the period of recovery from these illnesses. The major finding of this report was that the simple correlation coefficients relating duration of hospital stay to each individual pollutant concentration were statistically significant, though numerically small. Unfortunately, the linear regression coefficients, which could proxy for toxicity weights, were not published and could not be obtained. However, by dividing each correlation coefficient in Sterling et al. (1967) by the sample standard geometric deviation of the corresponding pollutant concentration, the relative values of the linear regression coefficients were estimated (see Kohn and Burlingame, 1971). These weights are listed in column 2 of table 5.5. There is some question (see Lin, Kohn, and Burlingame, 1972) as to whether arithmetic, rather than geometric, sample standard deviations should have been used to estimate the toxicity weights. Furthermore, these toxicity weights are based on length of hospital stay and reflect the short term effects of pollution and not the long term effects. However, a comparison of the toxicity weights used in this study with those developed independently by Zerbe and Croke (1975, p. 106) provides some verification. For purposes of comparison, the weights for the gaseous pollutants in column 2 of table 5.5 are converted from a part per million to a microgram per cubic meter basis. Each of the five weights are then divided by the toxicity weight for introgen oxides. The resulting figures are surprisingly close in relative magnitude to those of Zerbe and Croke, which appear in column 4 of the table.

Table 5.5

Comparison of toxicity weights

(1) Pollutant	(2) Relative toxicity weights based on data in Sterling et al.[a]	(3) Relative toxicity weights converted to microgram per cubic meter units and expressed as a ratio of the weight for nitrogen oxides[b]	(4) Relative toxicity weights from Zerbe and Croke[c]
Carbon monoxide	.10	.020	.055
Hydrocarbons	.16	.055	.419
Nitrogen oxides	6.8	1.00	1.00
Sulfur dioxide	48.0	4.15	3.86
Particulates	.012	2.81	2.20

[a]These denote the relative toxicity per part per million concentration of the gaseous pollutants and per microgram per cubic meter of particulates.
[b]The conversion factors in note 2 of chapter 4 are used here.
[c]Zerbe and Croke (1975, p. 106).

Table 5.6
Efficient standards for a given abatement budget from Model VIII

(1) Pollutant	(2) Given standards	(3) Relative toxicity weights[a]	(4) Efficient standards for an abatement budget of $46,525,242
Carbon monoxide	5.0 ppm	.10	5.8 ppm
Hydrocarbons	3.1 ppm	.16	3.3 ppm
Nitrogen oxides	.069	6.8	.080 ppm
Sulfur dioxide	.02 ppm	48.0	.14 ppm
Particulates	75 $\mu g/m^3$	.012	67 $\mu g/m^3$

[a]The relative toxicity weights are based on data in Sterling, Pollack, and Phair (1967).

The objective function for Model VIII contains the toxicity weights in column 3 of table 5.6 and is as follows:

$$Q = (.10)q^c + (.16)q^h + (6.8)q^n$$
$$+ (48.0)q^s + (.012)q^p. \tag{5.14}$$

This linear function is minimized subject to a budget constraint of $Z^* = $46,525,242$, which is the total cost from the solution of Model III. The results of this benefit-effectiveness model are contained in table 5.6. A comparison of the given standards in column 2 with the efficient standards in column 4 indicates that for carbon monoxide, hydrocarbons, and nitrogen oxides, the given standards are too stringent, while for sulfur dioxide and particulates the given standards are comparatively lax. To the extent that the given standards do approximate the legal air quality goals for the St. Louis airshed, these results imply that there has been an overemphasis on the first three pollutants, which are essentially automotive pollutants (see table 2.2), and insufficient abatement of the last two pollutants, which primarily originate from stationary sources. This result is consistent with findings of Kahn et al. (1974) with respect to carbon monoxide.

The linear pollution index is a useful adjunct of the Linear Programming Model.[8] We next consider the application of nonlinear index systems.

Benefit-Effectiveness and Nonlinear Pollution Index Systems

It is possible to express a nonlinear index system in piecewise linear form and thereby implement a model comparable to (5.13) above. However, this is a complicated procedure. An alternative version of benefit-effectiveness analysis facilitates the use of nonlinear pollutant indices. In this approach, the marginal rate of transformation, π^i/π^j, is compared to the marginal rate of damage. If a trade-off can be made that reduces damage without increasing the total cost of abatement, a shift in the indicated direction is economically efficient. This procedure is illustrated using the indices of Larsen and Green.

The Larsen (1970) index was derived from statistical studies of the death rate during air pollution episodes. This index, which relates sulfur dioxide and particulate concentrations to excess deaths, is

$$Q^L = (.6)(q^s)(q^p). \tag{5.15}$$

The marginal rate of damage between particulates and sulfur dioxide concentrations evaluated at the legal concentrations for the St. Louis airshed (.02 ppm and 75 $\mu g/m^3$) is

$$-dq^p/dq^s = 75/.02 = 3750. \tag{5.16}$$

In terms of a relatively small trade-off, this means that 3.75 $\mu g/m^3$ of suspended particulate matter is equivalent in toxicity to .001 ppm of sulfur dioxide.

The marginal rate of transformation between the two pollutants is the ratio of the shadow prices, π^s/π^p, from the solution of Model III:

$$-dq^p/dq^s = \pi^s/\pi^p \cong \$439,100,000/\$239,100 \cong 1840. \tag{5.17}$$

Thus, the sulfur dioxide concentration can be increased (decreased) by .001 ppm and the particulate concentration decreased (increased) by 1.84 $\mu g/m^3$ without any change in the total cost of abatement.

The marginal rate of damage is compared to the marginal rate of transformation in figure 5.6. It is clear that a shift rightward along ZZ' toward higher particulate concentrations and lower sulfur dioxide concentrations would achieve a lower valued, more desirable Q^L index level.

Green (1966) has developed an index for sulfur dioxide and particulate concentrations, which is also based on the observed impact of these pollutants on the death rate. This index, converted to the same units as $\mathbf{q}$ in the Linear Programming Model, is

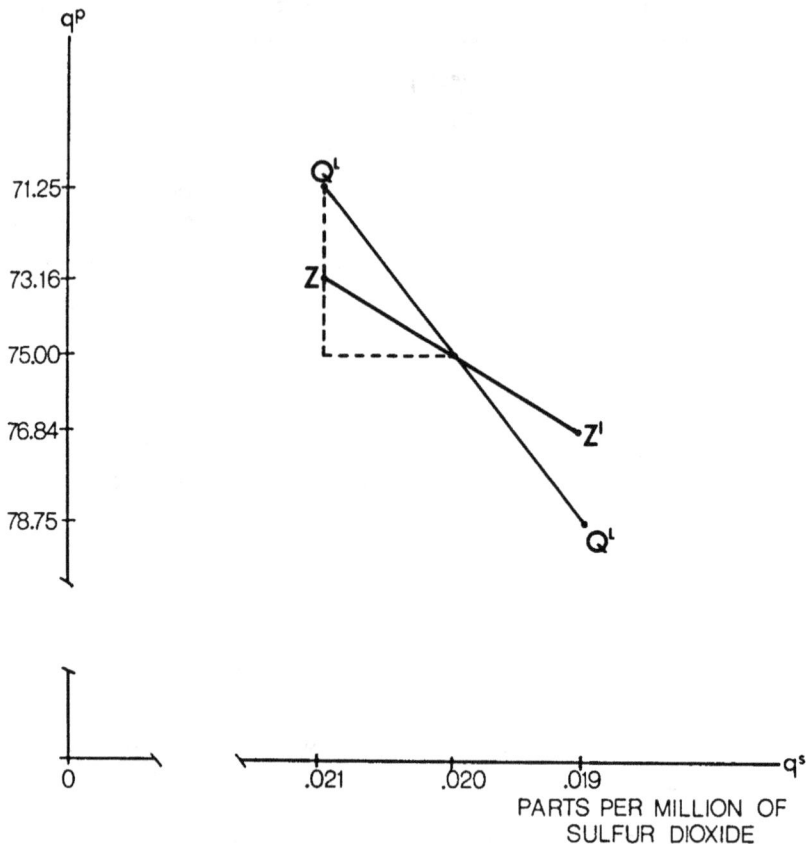

MICROGRAMS
PER CUBIC METER
OF PARTICULATES

Figure 5.6
A comparison of the marginal rate of substitution with the marginal rate of
transformation between two pollutant concentrations

$$Q = 84(q^s)^{.431} + 26.6(q^p/187.5)^{.576}. \qquad (5.18)$$

The marginal rate of substitution is

$$-dq^p/dq^s \simeq 48.16(q^p)^{.424}/(q^s)^{.569}. \qquad (5.19)$$

Evaluated at the legal concentrations, this is equivalent to a damage trade-off of approximately 2.78 $\mu g/m^3$ to .001 ppm. A comparison with the rate of transformation indicates, as with the Larsen index, that a trade-off in favor of lower sulfur dioxide concentrations is economically efficient.

The benefit-effective concentrations are those in which the marginal rate of substitution satisfies an inequality analogous to (5.11) above. This is illustrated in table 5.7. The numbers in columns 2, 3, and 4 of table 5.7 were obtained with a version of Model VIII in which the concentrations of carbon monoxide, hydrocarbons, and nitrogen oxides were fixed at the legal standards and Z was set equal to $\$46,525,242$. By minimizing q^p for a sequence of values of q^s, and using RANGE analysis (a subroutine of the MPS/360 program) to determine the extreme values of q^s for which the resulting shadow price of q^p remained constant, contiguous facets of the q^p, p^s frontier were determined. Table 5.7 includes a list of extreme points (a, b, c, etc.) in the neighborhood of the given standards.

The legal air quality goals for sulfur dioxide and particulate matter, $q^s =$ .02 ppm and $q^p = 75$ $\mu g/m^3$, lie on the facet bc. The rate of transformation on this particular facet is 1860 micrograms per cubic meter of particulates per part per million of sulfur dioxide.[9]

The benefit-effective concentrations based on the Larsen index are .0183 ppm and 80.17 $\mu g/m^3$. Based on the Green index, they are .0196 ppm and 75.75 $\mu g/m^3$. This analysis is, of course, based on the assumption that the technology of abatement and the damage relationship are correctly represented. Although the impact of pollution on the death rate is based on years of exposure, it is assumed here that the long run relative effect of each pollutant is proportional to the short run impact on which the Larsen and Green indices are based. It is of interest that the optimal sets of concentrations based on the Larsen and Green indices are fairly close to the legal standards in the St. Louis airshed.[10] If this were not the case, the benefits of the given abatement budget could be increased by altering the legal standards.

Table 5.7

Vertices and facets of the production-possibility frontier for annual average concentrations of sulfur dioxide and particulates, rates of transformation, and ratios of toxicities[a]

(1) Vertices and facets	(2) Sulfur dioxide concentration (ppm)	(3) Particulate concentration ($\mu g/m^3$)	(4) Rate of transformation $\left(\dfrac{-\mu g/m^3}{ppm}\right)$	(5) Larsen ratio[b] $\left(\dfrac{-\mu g/m^3}{ppm}\right)$	(6) Green ratio[c] $\left(\dfrac{-\mu g/m^3}{ppm}\right)$
a	.0208	73.65		3541	2700
ab			1671		
b	.0201	74.82		3722	2772
bc			**1860**		
c	**.0196**	**75.75**		**3865**	**2827**
cd			**3100**		
d	.0192	76.99		4010	2880
de			3233		
e	.0189	77.96		4125	2921
ef			3433		
f	.0186	78.99		4247	2964
fg			3850		
g	.0184	79.76		4335	2995
gh			**4100**		
h	**.0183**	**80.17**		**4381**	3011
hi			**4700**		
i	.0182	80.64		4431	3027
ij			5300		
j	.0180	81.70		4539	3064

[a]Benefit-effective concentrations corresponding to each pollutant index are shown in boldface.
[b]Larsen (1970).
[c]Green (1966).

Corner Optima and Nonconvexity

It can be seen in table 5.7 that the rate of transformation in column 4 increases as the particulate concentration in column 3 increases. The fact that the marginal rates of toxicity based on the Larsen and Green indices also increase with higher particulate concentrations indicates that the nonlinear pollution index isoquants and the production-possibility frontier are all concave to the origin. The concavity in the nonlinear pollution indices implies that if there are two sets of concentrations, such as (.014 ppm; 80 μg/m^3) and (.016 ppm; 70 μg/m^3), which yield the same index level (in this case, $Q^L = .672$, according to the Larsen index), then a linear combination of the two sets, such as (.015 ppm; 75 μg/m^3), will yield a higher and less desirable level for that same index. This concavity might be the consequence of synergism between the pollutants. This would be the case, for example, if the presence of particulate matter accelerated the conversion of sulfur dioxide to more noxious sulfuric acid, or if the adsorption of sulfur dioxide onto small particles caused the gas to be inhaled more deeply into the lungs. There is also a multiple stress theory in biology that would explain the concavity, even if there were no synergism.

When both the production-possibility frontier and the pollution index isoquants are concave, there is a potential nonconvexity in the Benefit-Effectiveness model. The traditional economic problem with nonconvexity relates to the use of Pigouvian fees. When there is nonconvexity, there is the possibility that pollution fees would promote a competitive market equilibrium that is a local optimum but not a global optimum (see Baumol and Oates, 1975, pp. 122–123). However, another problem posed by nonconvexity would exist if regulatory agencies sought to achieve benefit-effective concentrations and knew only the marginal rate of substitution (that is, the relative toxicity) of any pair of pollutant concentrations for any given level of air quality *but not the actual index values*. There would be no problem if at either of the extreme points of the production-possibility frontier between any two pollutants, the marginal rate of substitution, $-dq^i/dq^j$, were less than the marginal rate of transformation, π^j/π^i, where j is the pollutant whose concentration is lowest at the extreme point. This means that the index isoquants that are tangent to the frontier at each corner lie to the left of and below the end facets. In this case, which is illustrated in figure 5.7, the optimal solution must lie on an internal vertex. A comparison of the marginal

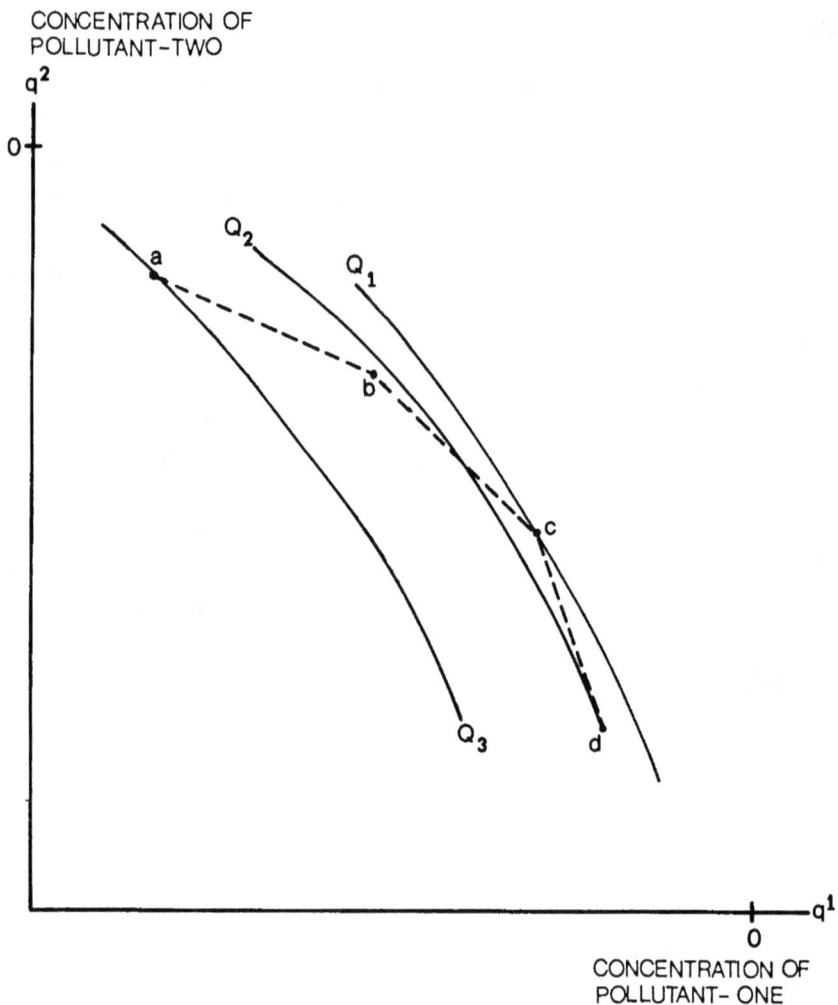

CONCENTRATION OF
POLLUTANT-TWO

q^2

CONCENTRATION OF
POLLUTANT- ONE

q^1

Figure 5.7
The case in which an internal vertex must be optimal

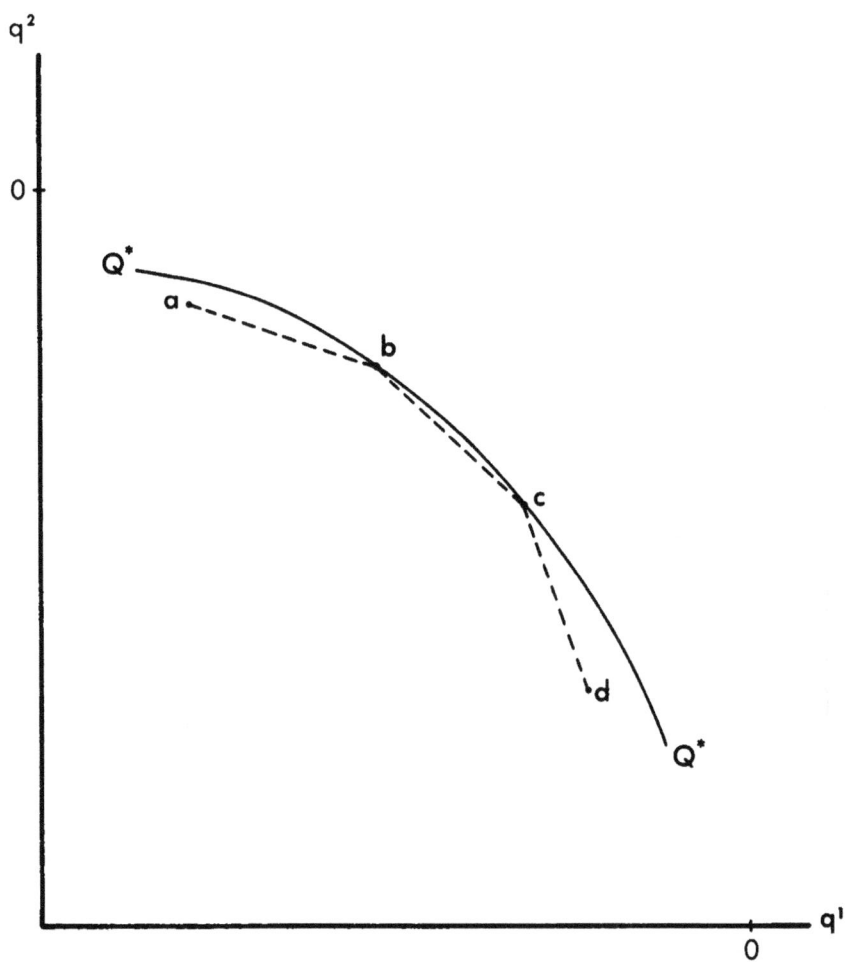

Figure 5.8
Multiple optima

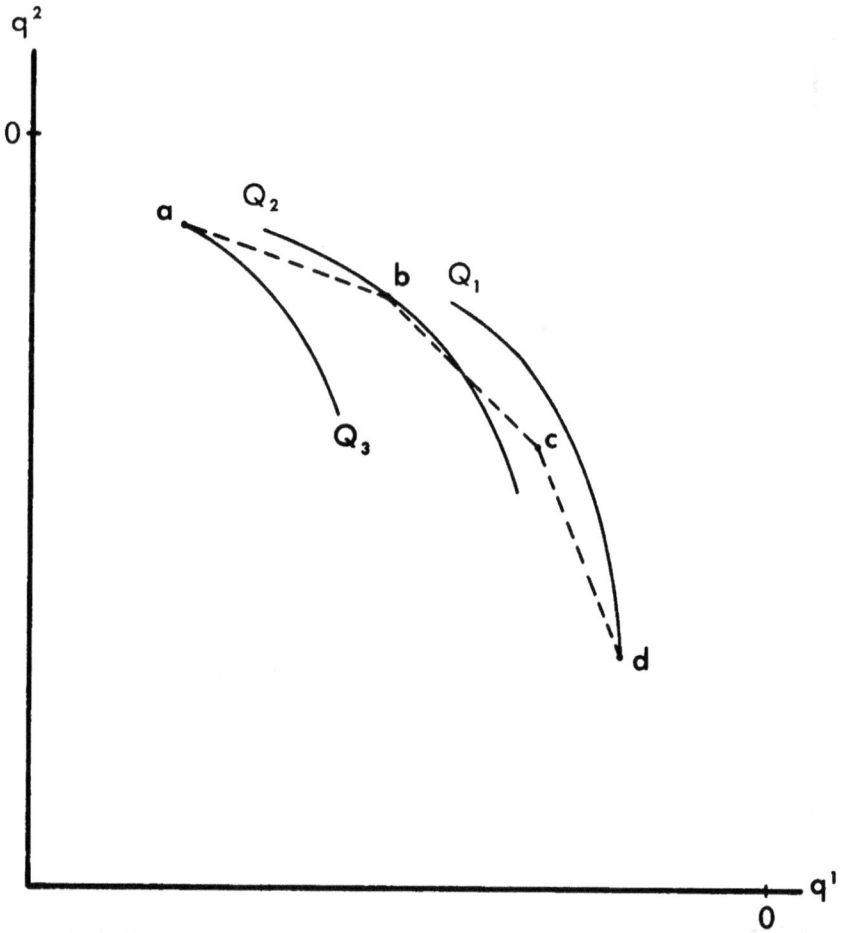

Figure 5.9
A corner optimum in which first-order conditions are also satisfied at an internal vertex

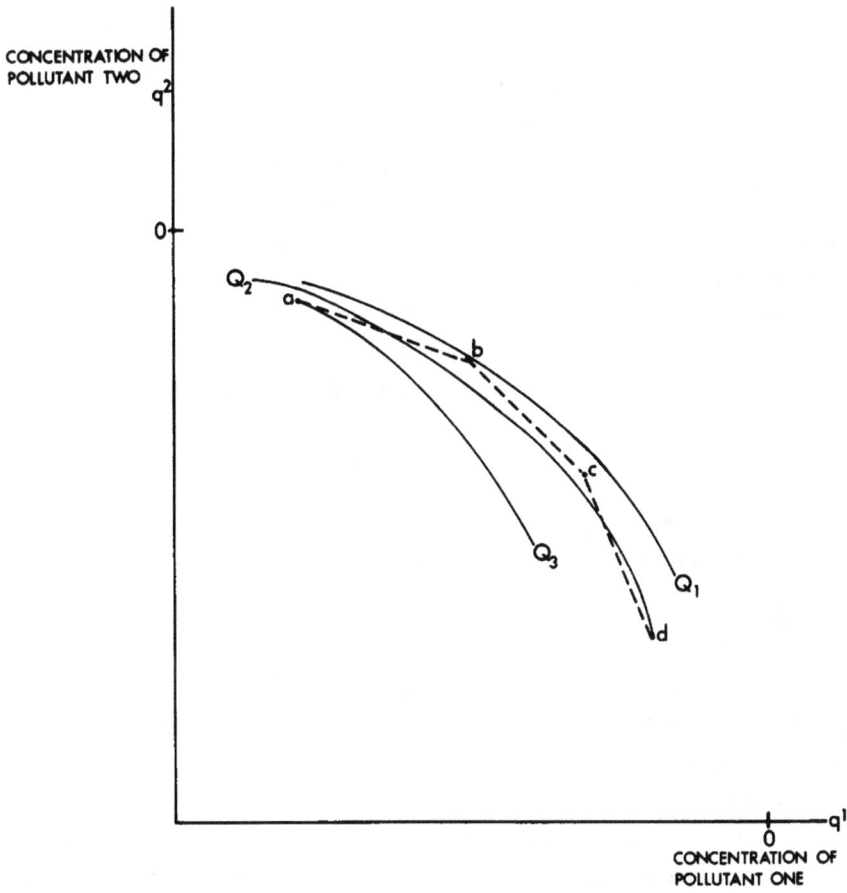

Figure 5.10
A vertex optimum in which first-order conditions are also satisfied at a corner

rate of substitution at each vertex with the rates of transformation on the adjacent facets, as in table 5.7, will lead to the global optimum. The possibility does exist that there will be multiple optima, as in figure 5.8; but these would be equally desirable.

When the marginal rate of substitution at one or both corners of the production-possibility frontier exceeds the rate of transformation on the adjacent facet, as in figures 5.9 and 5.10, there may be either a corner solution as in the former or an internal vertex solution as in the latter.[11] In either case, the first order condition (such as 5.10 above) for an optimum is satisfied both at a corner (or corners) and at a vertex. Unless the index values Q were known, there would be no way to distinguish the global optimum from the local optimum. Because of the problem of nonconvexity, it may therefore be necessary for regulatory agencies to know index values as well as marginal rates of substitution between pollutant concentrations.

Divisibilities in the Control Method Solution

The solution of Model I includes five divisibilities in the optimal set $\mathbf{x}_i^*$, one for each pollutant requirement that is binding.[12] It follows from the discussion of the solution of (5.6) that there would be a single divisibility in the solution of Model VIII, in which benefit-effective concentrations of five pollutants are determined. Such, indeed, is the case.

In Model VII, in which the dollar value of pollution damage plus total abatement costs is minimized, there are no divisibilities in the activity solution. This is the case because the damage function is linear. The production-possibility frontier is analogous to the simplex abcd in figure 5.11, which is a tetrahedron in the space defined by the three axes. It is apparent that the hyperplane $w^1 q^1 + w^2 q^2 + z$ is most likely to be tangent to the simplex at one of the vertex points.[13] In the more general case, such as that represented by (5.6) above, with conventional utility functions, there will be a single divisibility in the activity solution.[14]

In chapters 1 and 3 it was noted that divisibilities in the optimal control method solution are difficult to achieve by regulatory directives and impossible to promote by Pigouvian fees. If the harmful effects of air pollution can be measured by index systems, however, it follows from the discussions in this chapter that optimal pollutant concentrations will be largely associated with integer abatement activity levels.

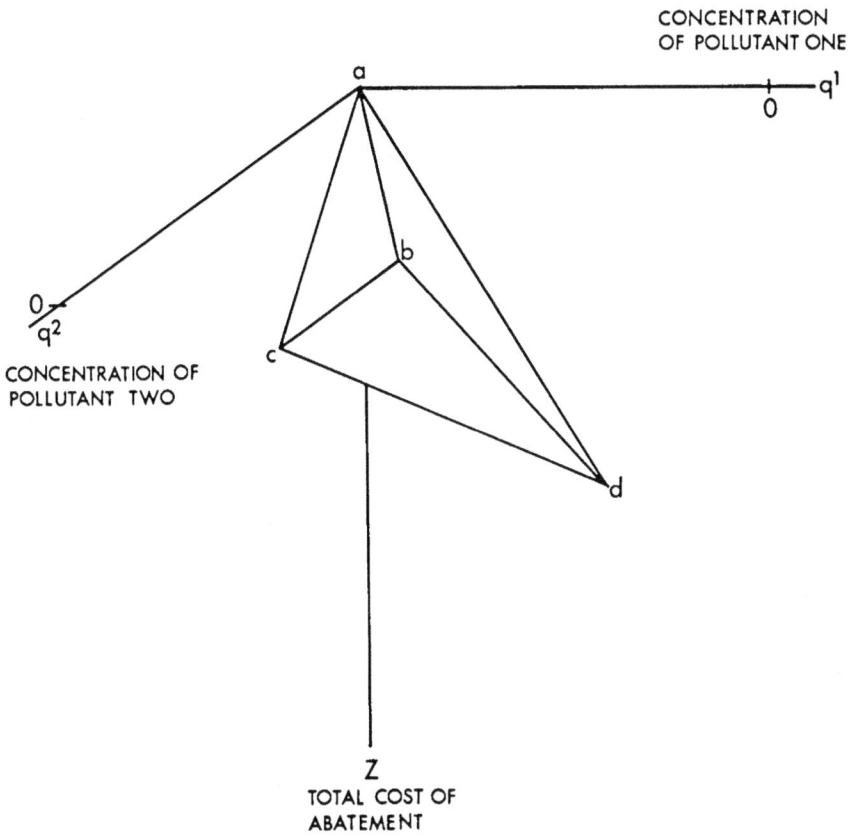

Figure 5.11
Production-possibility frontier between two pollutant concentrations and the total cost of abatement

Chapter Summary

In this chapter we have related data on pollution damage to the cost and technology of abatement. This was accomplished through the application of benefit-cost analysis, which, given certain restrictions, can be applied to the multipollutant case in a general equilibrium context. The shadow prices of the pollutant standards in the Linear Programming Model have a special significance; if the standards are optimal, the shadow prices are a measure of the marginal benefits of abatement. The pollutant shadow prices from Models I and III compared favorably with the damage estimates of other investigators. A linear damage function was formulated, using the shadow prices, and it was shown that optimal pollutant concentrations change as the cost of inputs for abatement rise or fall.

The conditions defining economic efficiency in the multipollutant case are exceedingly complex. These would be greatly simplified if air quality could be represented by a single index in which the individual pollutant concentrations are appropriately weighted. Furthermore, the optimal control method solution would contain no more than one divisibility. An index relating length of hospital stay to pollutant concentrations was developed and used in a benefit-effectiveness version of the Linear Programming Model. The results of this model suggest that for the given abatement budget there should be some shift in emphasis from the automobile-related pollutants to sulfur dioxide and particulate abatement.[15]

The air pollution literature includes several index functions for sulfur dioxide and particulates. When these functions are related to the technology of abatement in the St. Louis airshed, the optimal concentrations for these pollutants are found to be very close to the legal standards. In this analysis it is observed that the nonlinear index functions for sulfur dioxide and particulates are concave, as is the production-possibility frontier. Thus, there is a potential nonconvexity in the Benefit-Cost model. This nonconvexity could be a problem if regulatory agencies sought to establish benefit-effective concentrations and knew only the relative toxicities of pollutants for any given set of concentrations. If they could not compare or order index values, they would be unable to distinguish local from global optima. This chapter demonstrates the usefulness of an air quality index based on concentrations of the individual pollutants and illustrates that the index must at least be ordinal.

APPENDIX: NUMERICAL EXAMPLE

A numerical example of a pollution index is

$$Q = .8q^1 + .2q^2. \tag{5.20}$$

The conditions for economic efficiency can be illustrated for a simple economy consisting of two households whose utility functions are

$$U^1 = 170 \ln Y_1 + 11 \ln (5200 - Q),$$
$$U^2 = 5 \ln Y_2 + \ln (5200 - Q). \tag{5.21}$$

Given the same production-abatement technology and quantity of resources as in the appendix of chapter 1, the following would be a Pareto optimal allocation:

$$Y_1 = 920, \; Y_2 = 870,$$
$$q^1 = 34{,}000, \; q^2 = 2475, \; Q = 3215, \tag{5.22}$$
$$x_{1c} = 475, \; x_{1e} = 25, \; z = 210.$$

This allocation satisfies condition (5.7), because

$$- U_Q^1/U_Y^1 - U_Q^2/U_Y^2 = \frac{(11)(920)}{(1985)(170)} + \frac{(870)}{(1985)(5)} = \frac{2}{17}$$

$$= \frac{c_{1e} - c_{1c}}{.8(e_{1c}^1 - e_{1e}^1) + .2(e_{1c}^2 - e_{1e}^2)} = \frac{1.8 - 1.4}{.8(7 - 3) + .2(5 - 4)} = \frac{2}{17}.$$

An alternative version of this example is the minimization of pollution damage, given an abatement budget of $Z = 210$. This benefit-effectiveness model has the form

Minimize $Q = .8q^1 + .2q^2$
subject to
$$q^1 - 10x_{1a} - 8x_{1b} - 7_{1c} - 7x_{1d} - 3x_{1e} = 0,$$
$$q^2 - 12x_{1a} - 10x_{1b} - 5x_{1c} - 3x_{1d} - 4x_{1e} = 0, \tag{5.24}$$
$$.1x_{1a} + .2_{1b} + .4x_{1c} + .45x_{1d} + .8x_{1e} = 210,$$
$$x_{1a} + x_{1b} + x_{1c} + x_{1d} + x_{1e} = 500.$$

The constraints in (5.24) define a faceted isoquant for $Z = 210$ comparable to the isoquant $Z = 130$ in figure 1.6, except that the new isoquant is northeast of point c and therefore consists of four line segments. The vertices of this isoquant are (18500/6; 23400/6), (3400; 2475), (3500; 2100), (3560; 1920), (127000/35; 66000/35). The index minimizing combination of concentrations is defined by the tangency of the line $Q = .8q^1 + .2q^2$ at the vertex $(3400; 2475)$, which is where the isoquant $Z = 210$ crosses the line segment ce in figure 1.6. The slope of the index line is $w^1/w^2 = 4$, while the slopes (see 1.58) of the two line segments that join at $(3400; 2475)$ are $.09375/.025 = 3.75$ and $.09474/.02105 \cong 4.5$. Thus, inequalities corresponding to (5.11) are satisfied.

6

THE FEEDBACK OF ABATEMENT ACTIVITY ON POLLUTION SOURCE LEVELS

Pollution originates in the production and consumption of various goods. In the Pure Abatement model it is assumed that the sources of pollution are intermediate activities which, by virtue of a special assumption on technology, are fixed in quantity. Accordingly, there are no substitutions in production and consumption that would diminish these source levels, and pollution control is accomplished entirely by technological abatement.

In this chapter we depart from the Pure Abatement model and allow for the feedback of abatement activities on pollution source levels. This analysis is facilitated by distinguishing a vector of final goods and assuming that the pollution source levels are linearly related to the outputs of these goods. There are two distinct kinds of feedbacks that are examined. The first is a consequence of the incremental costs of abatement that cause prices of the final goods to increase and quantities demanded to decrease. Accordingly, the pollution source levels, hitherto fixed, would decline. The second feedback incorporates the increased demand for goods as inputs to abatement, and as a consequence, increases in pollution source levels. Thus, the two feedbacks have opposite effects on the total cost of abatement: in the first case, total cost is decreased; in the second, it is increased.

Output of Goods

The format of Model I is

Minimize $\mathbf{Cx}$

subject to $\mathbf{ux} = \mathbf{s}$,

$$\mathbf{ex} \leqslant \hat{\mathbf{f}}, \tag{6.1}$$

$$\mathbf{x} \geqslant 0.$$

The basic assumption of the Pure Abatement model is that the elements of the $\mathbf{s}$ vector (quantities such as gallons of diesel fuel consumed in trucks, tons of solid waste incinerated, etc.) are proportional to labor input, that is,

$$\begin{bmatrix} s_1 \\ s_2 \\ s_3 \\ \vdots \\ s_m \end{bmatrix} = \begin{bmatrix} \alpha_1 \\ \alpha_2 \\ \alpha_3 \\ \vdots \\ \alpha_m \end{bmatrix} R, \tag{6.2}$$

where R is a scalar denoting the fixed quantity of the single input. In this model the s_i are constant and there is no interaction between the $\mathbf{x}$ and $\mathbf{s}$ vectors.

We now incorporate in the model a $t \times 1$ vector of goods $\mathbf{y}$. The elements of this column vector denote quantities of goods produced, such as tons of steel, kilowatt hours of electricity, barrels of cement, etc. We assume that there is a technical relationship between pollution source levels and outputs, such that

$$\begin{bmatrix} s_1 \\ s_2 \\ s_3 \\ \vdots \\ s_m \end{bmatrix} = \begin{bmatrix} a_{11} & a_{12} \cdots a_{1t} \\ a_{21} & a_{22} \cdots a_{2t} \\ a_{31} & a_{32} \cdots a_{3t} \\ \vdots & \vdots \quad \vdots \\ a_{m1} & a_{m2} \cdots a_{mt} \end{bmatrix} \begin{bmatrix} y_1 \\ y_2 \\ y_3 \\ \vdots \\ y_t \end{bmatrix} \tag{6.3}$$

Accordingly, each source level is linearly related to the set of outputs. For example,

$$s_3 = a_{31} y_1 + a_{32} y_2 + \cdots + a_{3t} y_t, \tag{6.4}$$

where a_{ik} is the required input of pollution source i per unit of good k. In matrix notation the constraint on activity levels is now

$$\mathbf{ux} = \mathbf{ay}, \tag{6.5}$$

where $\mathbf{a}$ is the $m \times t$ matrix whose typical element is a_{ik}, and $\mathbf{u}$ is the $m \times n$ distributive matrix whose element u_{ij} is one when the jth abatement process is defined for the ith pollution source and zero otherwise.[1]

The Linear Programming Model can be expressed in terms of outputs, as follows:

Minimize $\mathbf{Cx}$

subject to $\mathbf{ux} = \mathbf{ay}$,

$$\mathbf{ex} \leqslant \hat{\mathbf{f}}, \tag{6.6}$$

$$\mathbf{x} \geqslant 0.$$

This formulation of the model implies that total emission flows can be related to outputs of goods, rather than categories of pollution sources as in tables 2.2 and 3.5. A breakdown of goods by broad output categories is contained in table 6.1. There are eighteen classifications in this table, including intermediate goods, final goods, services, and combinations of goods and services. The output levels of the eighteen categories in table 6.1 could be measured in such units as miles of automobile transportation, Btu of heat from residential furnaces, kilowatt hours of electricity, tons of steel, etc. These output levels would be elements of the vector $\mathbf{y}$.

The data in table 6.1 were estimated by allocating the 97 source magnitudes in table 2.1 to the appropriate economic activities. Certain sources of pollution, such as industrial furnaces and incinerators, were distributed among at least eight of the industrial activities in table 6.1. Observe that the classifications in table 2.2 are polluting activities, whereas those in table 6.1 are essentially economic activities. The total emission flows are necessarily the same in both tables.

The emission flows in table 6.1 are predicated on base-year abatement activities. Given a least-cost set of control methods for achieving the total allowable emission flows, and assuming no feedbacks of the control method solution on the levels of pollution-related outputs, the breakdown of emission flows in 1975 would be that shown in table 6.2. Because they are derived from the same data, the total flows in table 6.2 are, of course, equal to those in table 3.5.

It should be explained that the categorization of economic sectors in tables 6.1 and 6.2 was done after the research on feedbacks, described in this chapter, had been completed. Thus, the research in this chapter is based on different classifications of outputs. However, the basic principles involved are the same.

Interaction of the Abatement Activity Vector and Output Levels

There are likely to be feedbacks of the optimal solution $\mathbf{x}^*$ on the vector of final outputs $\mathbf{y}$. In that case,

$$\mathbf{y} = \mathbf{y}^0 + \Delta\mathbf{y}, \tag{6.7}$$

where $\mathbf{y}^0$ is the vector of projected outputs in the absence of regulation, and $\Delta\mathbf{y}$ is the vector of output changes resulting from implementation of *alternative* abatement activities. Note that the emphasis is on alternative activities;

Table 6.1

Projected emission flows in the St. Louis airshed in 1975 in the absence of a regulatory program (millions of pounds)

(1) Economic activities	(2) Carbon monoxide	(3) Hydro- carbons
Consumption Activities		
Automobile driving	3,309	700
Residential heating	23	54
Miscellaneous	n	16
Service Activities		
Bus transportation	n	1
Air transportation	13	3
Dry cleaning	n	10
Combustion of residential refuse	48	147
Commercial and institutional activities	4	17
Industrial Activities		
Electric power generation	3	6
Oil refining	750	254
Sulfuric acid production	n	1
All other chemical industries	4	63
Cement production	1	2
Steel production	1	5
Iron and steel foundries	1	4
Lead smelting	n	n
Truck, barge, rail transportation	4	13
Miscellaneous industrial activities	41	222
Totals	4,202	1,518

Note: n indicates less than 500,000 pounds.

(4) Nitrogen oxides	(5) Sulfur dioxide	(6) Particulates
128	10	13
27	61	26
n	n	9
2	n	1
4	n	5
n	n	n
1	2	32
6	16	7
136	755	34
16	54	6
1	68	1
28	66	20
1	n	31
7	54	29
2	2	3
n	174	n
15	3	8
42	124	75
416	1,389	300

Table 6.2

Allowable emission flows in the St. Louis airshed in 1975 in compliance with a least-cost set of regulations (millions of pounds)

(1) Economic activities	(2) Carbon monoxide	(3) Hydro- carbons
Consumption Activities		
Automobile driving	2,229	477
Residential heating	n	53
Miscellaneous	n	16
Service Activities		
Bus transportation	n	1
Air transportation	13	3
Dry cleaning	n	10
Combustion of residential refuse	8	10
Commercial and institutional activities	3	17
Industrial Activities		
Electric power generation	3	6
Oil refining	39	136
Sulfuric acid production	n	1
All other chemical industries	3	61
Cement production	n	2
Steel production	1	5
Iron and steel foundries	n	4
Lead smelting	n	n
Truck, barge, rail transportation	4	12
Miscellaneous industrial activities	32	180
Totals	2,335	994

Note: n indicates less than 500,000 pounds.

(4) Nitrogen oxides	(5) Sulfur dioxide	(6) Particulates
72	10	13
25	6	4
n	n	9
2	n	1
4	n	5
n	n	n
1	1	9
6	10	2
94	99	3
16	49	5
1	7	1
21	47	4
1	n	22
6	47	5
2	2	3
n	35	n
15	3	8
38	84	42
304	400	136

the economic impact of the base-year abatement activities is already incorporated in the production levels $\mathbf{y}^0$. It is the feedback of the alternative activities that affects the solution of the Linear Programming Model.

To retain the linear format, we assume that the change in output of the kth good is a summation of changes caused by alternative abatement activities

$$\Delta y_k = \Delta y_k^{1b} + \Delta y_k^{1c} + \Delta y_k^{2b} + \cdots \tag{6.8}$$

and that the latter are proportional to the corresponding activity levels. The constant of proportionality is expressed by the coefficient β_k^j, which denotes the change in the output of good k as a consequence of a unit of activity of process j. For example,

$$\begin{aligned}
\Delta y_k^{1b} &= \beta_k^{1b} x_{1b}, \\
\Delta y_k^{1c} &= \beta_k^{1c} x_{1c}, \\
\Delta y_k^{2b} &= \beta_k^{2b} x_{2b}.
\end{aligned} \tag{6.9}$$

The revised Linear Programming Model is

Minimize $\mathbf{Cx}$
subject to $[\mathbf{u} - \mathbf{a}\beta]\mathbf{x} = \mathbf{a}\,\mathbf{y}^0,$
$$\begin{aligned}
\mathbf{ex} &\leqslant \hat{\mathbf{f}}, \\
\mathbf{x} &\geqslant 0,
\end{aligned} \tag{6.10}$$

where β is the $t \times n$ matrix whose typical element is β_k^j. (Note that the β matrix is generalized by including the zero coefficients for the base-year activities.) In this version of the Linear Programming Model, there is no interaction between the decision variables x_j and the right-hand-side parameters.[2]

Two types of feedbacks, each resulting in a different β matrix, are examined in this chapter. Observe that there is no constraint on resources in (6.10). Abatement can result either in an influx or an exit of resources from the airshed.

Feedback of the Costs of Abatement on Quantities of Goods Demanded

The adoption of alternative abatement activities will increase the selling prices of goods. This impact is incorporated in Model IX, an expanded version of Model I, in which the increase in selling prices reduce quantities demanded. In this analysis we are interested in the *incremental* costs of abatement for each activity, and we define a new cost of abatement vector, $\hat{C}$, in which the element C^j is zero for a base-year activity and is the incremental cost of abatement for alternative activities. For example,

$$C^{2a} = C_{2a} - C_{2a} = 0,$$
$$C^{2b} = C_{2b} - C_{2a}, \qquad\qquad\qquad (6.11)$$
$$C^{3d} = C_{3d} - C_{3a}.$$

We have seen in chapters 2 and 3 that certain components of the cost of abatement are costs that would not affect the accounting decisions of firms and households. For example, the cost of converting furnaces from coal to natural gas includes a scarcity premium that reflects the relative scarcity of natural gas, but which is not included in the regulated price. In this analysis, we shall assume that the elements C^j of the vector $\hat{C}$ represent private costs only.

Associated with each good k is a unit production cost, c_k, which includes the costs associated with the base-year level of abatement. The implementation of process j to control the pollution from source i will increase the total cost of producing good k by $a_{ik}C^j$. This implies that all of source i is controlled by process j, that is, $x_j = s_i$. To allow for divisible solutions, the increase in total cost of good k is represented by $a_{ik}C^j(x_j/s_i)$. The proportional increase in the production cost of good k as a consequence of abatement activity j, which is defined for source i, is therefore $a_{ik}C^j(x_j/s_i)/c_k$.

We shall assume that prices of the goods are proportional to their production-abatement cost. The decrease Δy_k^j, in the quantity of good k demanded, as a consequence of abatement activity j, is based on the formula for the price elasticity of demand, as follows:

$$\varepsilon_k = \frac{\Delta y_k^j / y_k^0}{a_{ik}C^j x_j/(s_i c_k)}. \qquad\qquad\qquad (6.12)$$

Therefore,

$$\Delta y_k^j = \left(\frac{\varepsilon_k a_{ik}C^j y_k^0}{c_k s_i}\right)x_j. \qquad\qquad\qquad (6.13)$$

The coefficient of x_j in (6.13) corresponds to β_k^j in (6.10) above. This coefficient, which we shall call v_{kj}, is formally defined as follows. For each abatement activity j there is a feedback coefficient v_{kj} on each of the t-

outputs. This coefficient denotes the change in output of good-k per unit of the jth activity. When the jth variable is the base-year activity, C^j is zero and therefore v_{kj} is zero, or when an abatement activity and a particular output are not related, the a_{ik} is zero, as is the corresponding v_{kj}. Because the elasticities are negative, the coefficients v_{kj} are negative.

For simplicity, the value of s_i in (6.13) is taken at the projected preregulatory value. It follows that $a_{ik} = s_i/y_k$ for sources s_i that are proportional to a single output y_k. For sources that are a linear combination of several activities, $a_{ik}y_k/s_i$ is equal to the proportion P_{ik} of source level i used by industry k. Accordingly, (6.13) can be simplified, and v_{kj} becomes

$$v_{kj} = \frac{\varepsilon_k C^j P_{ik}}{c_k} \tag{6.14}$$

The price effects of alternative abatement activities are incorporated in Model IX by the constraint

$$[\mathbf{u} - \mathbf{av}]\mathbf{x} = \mathbf{a}\mathbf{y}^0, \tag{6.15}$$

where $\mathbf{v}$ is the $t \times n$ matrix of feedback coefficients. The product $a_{ik}v_{kj}$ is the reduction in source level i as a consequence of abatement activity j. For example, the reduction in source 4, thousands of gallons of diesel fuel burned in buses, as a consequence of abatement activity 4b, which represents the use of a less polluting diesel fuel, is derived from the following coefficients:[3]

$a_{4,k}$ = .000238 thousand gallons of diesel fuel per unit of bus output,
C^{4b} = \$14.36 per thousand gallons of diesel fuel,
ε_k = −1.2 for bus transportation, $\qquad$ (6.16)
$P_{4,k}$ = 1.0,
c_k = \$1.00 per one dollar unit of bus transportation (at preregulatory price).

The reduction in source level is .004 thousand gallons of diesel fuel per unit of activity 4b.

Another example is activity 20e, which represents the conversion of certain chain grate stokers from coal to natural gas. A number of different industries use this category of stoker (source number 20). The impact of abatement activity 20e on one of these industries, the Stone, Clay, and Glass Industry (which accounts for an estimated 12 percent of the coal burned in source number 20), and the reduction in source level as a consequence of that impact can be derived from the following coefficients:

$a_{20,k}$ = .0002 tons of coal burned in source number 20 per unit of output of the Stone, Clay, and Glass Industry;

ε_k = −.3 for the Stone, Clay, and Glass Industry; $\qquad$ (6.17)

C^{20e} = \$4.46 per ton of coal (exclusive of a scarcity premium of \$3.80);

$P_{20,k}$ = .12 of source 20 occurs in the Stone, Clay, and Glass Industry;

c_k = \$1.00 per one dollar unit of industry output (at preregulatory prices).

The reduction in source 20 as a consequence of the cost of activity 20e on the output of the Stone, Clay, and Glass Industry is .00003 tons of coal per unit of abatement activity.

The elements of the **v** matrix were calculated using price elasticities of demand taken from the economic literature. There is a conceptual problem here in that the derivation of the price elasticity of demand for a particular good is based on the assumption that the prices of all other goods remain constant. Thus, it is incorrect to incorporate a set of price elasticities of demand $\varepsilon_1, \varepsilon_2, \cdots, \varepsilon_l$, each predicated on a partial equilibrium analysis, into the context of a general equilibrium model. However, no other methodology for estimating the simultaneous effect of abatement costs was available. Nor were cross-elasticities of demand which imply *increases* in quantities demanded as a consequence of higher prices of substitute goods taken into account. Whereas the absence of cross-elasticities of demand overstates the price feedback on source levels, the evaluation of s_i in (6.13) at the preregulatory levels causes a bias in the opposite direction.

Before examining the optimal solution of Model IX, we shall evaluate the feedback effect of the solution $\mathbf{x}_1^*$ of Model I (see column (11) of table 2.1).[4] Allowing for the feedbacks of that solution and the estimated decreases in pollution source levels, there would be excess abatement of each pollutant, and the annual emission flows for the five pollutants would range from .1 to 1.0 percent below the allowable flows. Although the excess abatement is small, there is some potential for reducing the total cost of abatement by incorporating the feedback coefficients in the optimization model.

The Loss of Consumer Surplus

We noted in chapter 1 that the cost of abatement $\mathbf{Cx}^*$ understates the economic cost because there is no allowance for the loss of consumers' surplus (see figure 1.5). This loss was accounted for in Model IX by multiplying the increase in the price of each output by the decrease in quantity demanded (see Kohn, 1972b, p. 393). Unfortunately, this methodology overstates the loss of consumers' surplus because the area under the downward sloping demand curve, which represents the reduction in consumer surplus, is more likely to be a triangle than a rectangle.[5] To incorporate the estimated loss of consumers' surplus, the objective function in (6.10) is changed to

$$\text{Minimize } \mathbf{Cx} + \mathbf{1v\hat{C}x}, \tag{6.18}$$

where $\hat{\mathbf{C}}$ is an $n \times n$ diagonal matrix whose diagonal entry is C^j and $\mathbf{1}$ is a $1 \times t$ row vector of 1's.

The Model Incorporating the Feedback of the Costs of Abatement

The format of Model IX is

$$\text{Minimize } [\mathbf{C} + \mathbf{1v\hat{C}}]\mathbf{x}$$
$$\text{subject to} \tag{6.19}$$
$$[\mathbf{u} - \mathbf{av}]\mathbf{x} = \mathbf{ay}^0,$$
$$\mathbf{ex} \leqslant \hat{\mathbf{f}},$$
$$\mathbf{x} \geqslant 0.$$

The solution of this model includes $34.7 million in incremental control method costs (see table 6.3). Thus, the feedback effect in Model IX allows a saving of $.6 million in the direct costs of abatement. This is approximately 1.5 percent of the total cost. The loss of consumers' surplus corresponding to the optimal solution of Model IX is $.2 million. This, plus the $34.7 million above, gives a total of $34.9 million in incremental economic costs.[6]

Model IX may be viewed as an attempt to determine a least-cost set of abatement activities in a realistic general equilibrium context. The substitutions in consumption, which are assumed away in the Pure Abatement model, are anticipated. These reduce the total cost of technological abatement. The cost of pollution control is the opportunity cost of the resources allocated to abatement plus an estimate of the loss of consumers' surplus. The latter is approximately .5 percent of the former and is comparatively minor.

Induced Substitutions of Energy Inputs

The data for activity 20e in (6.17) indicate that the feedback effect (of converting travelling grate stokers from coal to natural gas) via the Stone,

Table 6.3

Total incremental cost of abatement for various versions of the Linear Programming Model

Description	Model number	Total incremental cost[a] (dollars)
The original model	I	35,337,282
Model with feedback of abatement costs on selling prices	IX	34,691,163
Model with feedback of abatement cost on selling prices and induced fuel substitutions	X	34,136,553
Model incorporating direct demand for inputs for abatement	XI	35,816,305
Model incorporating direct and indirect demand for inputs for abatement	XII	36,091,284

[a]Direct costs only; consumer surplus losses not included.

Clay and Glass Industry is insignificant. This is the case because pollution control of industrial furnaces is responsible for a very small proportional increase in total production costs. Similarly, abatement at the electric power plant causes only a small increase in production costs of industries. It is likely, however, that these cost increases would induce some fuel substitutions by producers. Model X, an extension of Model IX, incorporates estimated substitutions by industry of natural gas for both coal and electricity, in response to higher costs of these energy sources due to abatement. With Model X, the direct cost of abatement Cx^* declined by an additional \$.6 million per year as compared to the total cost from Model IX.

In the activity solution of Model X, the required tonnage of low sulfur coal is almost twice as great as that in Model I. This is the case because Model X incorporates elasticities of fuel and energy substitution. Control officials in the St. Louis area have acknowledged that part of the rationale for the low sulfur requirement on coal was to stimulate voluntary conversions to natural gas. The solution of Model X provides confirmation that this rationale was to some extent efficient. As was noted in chapter 3, however, it would be inefficient for purposes of air pollution control to convert pulverized coal furnaces to natural gas.

Inputs for Abatement

In the Pure Abatement model in chapter 1, some of the intermediate good is used as an input in its own production. The control of pollution from the production of this good requires additional input of the intermediate good itself. Accordingly, the required abatement is necessarily greater. It was Leontief (1970) who first observed that air pollution abatement generates demand for inputs, the production of which is polluting. Empirical work in this area was subsequently done by Miernyk and Sears (1974) and by Rose (1975).

With the Leontief feedback, there is a direct increase in the output of good k as a consequence of abatement activity j. This relationship is assumed to be

$$\Delta y_k^j = h_{kj} x_j, \tag{6.20}$$

where h_{kj} is the quantity of good k used as an input in implementing abatement activity j. Note that Δy_k^j is proportional to x_j, so that h_{kj} is representative of the general feedback coefficient β_k^j in (6.9) and (6.10).

In this extension of the Linear Programming Model, called Model XI, the **y** vector is chosen to conform to the sector classification in the input-output model of the St. Louis region developed by Liu (1968). To simplify the accounting, the inputs for abatement are traced to only six of these economic sectors, and the y_i are measured in million-dollar output units. The coefficients h_{kj} for three of the abatement activities are shown in table 6.4. The negative entries represent the value of coal replaced by natural gas and the labor and miscellaneous savings gained through conversion.[7] Inputs that are obtained from other sectors are included in table 6.4 under miscellaneous. Any feedback effect of these other inputs is not incorporated in the model. It is assumed that capital investments in control equipment are additional expenditures in the region and do not crowd out other investments; accordingly, no feedback is attributed to the opportunity cost of capital. Nor is there a feedback for the scarcity premium for natural gas.

Note that the unit costs of abatement in table 6.4 are the incremental costs; the demand for inputs associated with the base-year abatement activities are already accounted for in the output vector $\mathbf{y}^0$.

It was noted in table 3.1 that the incremental cost of air pollution abatement obtained with Model I is \$35.3 million for the year 1975. The portion of this cost that reflects purchases from the six specified economic sectors is the product $\mathbf{hx}_i^*$. These are entered in the first seven rows of table 6.5. It is

Table 6.4

Value of inputs per activity unit for selected abatement methods (dollars)

Abatement activity	Number 21e: tons of coal (burned by industry in chain grate stokers) replaced by natural gas	Number 41c: tons of coal burned in the Meramec power plant with the scrubbing process	Number 56b: tons of solid waste compacted for landfill instead of being burned
Incremental unit cost of abatement	8.17	1.20	11.16
Value of inputs from the chemical, petroleum, and rubber products sector	0	.22	.30
Value of inputs from the nonelectric machinery sector	0	.19	0
Value of inputs from the transportation equipment sector	0	0	1.50
Value of inputs from the mining sector	−5.27	0	0
Value of inputs from the utilities sector	11.27	.20	0
Value of inputs from the household sector	−1.57	.15	7.20
Miscellaneous unallocated inputs	−.06	.05	1.16
Opportunity cost of capital	0	.39	1.00
Scarcity premium for natural gas	3.80	0	0

these expenditures, totalling $24.0 million, that would have a feedback effect in Model XI. The remaining $11.3 million in costs would not.[8]

In Model XI, the appropriate **a** matrix is 97 × 23. For purposes of illustration the 41st column of that matrix is shown in table 6.6. In general, a polluting source is not related to every economic sector, and many of the a_{ik} coefficients are equal to zero. The coefficients a_{ik}, which denote units of source i per unit of good k, are derived from data for 1967 and are assumed to be constants.

The format of Model XI is as follows:

Minimize **Cx**
subject to $[\mathbf{u} - \mathbf{ah}]\mathbf{x} = \mathbf{ay}^0$,

$$\mathbf{ex} \leqslant \hat{\mathbf{f}},$$
$$\mathbf{x} \geqslant 0,$$

(6.21)

where **h** is a $t \times n$ matrix containing the elements h_{kj}. In this model we are ignoring the price effect of abatement on the output levels of goods and assuming that sufficient resources are attracted to the region to produce the inputs needed for abatement. The latter assumption is consistent with the "open economy" premise of input-output analysis.

The solution of Model XI yields a total cost of abatement that is $.5 million higher than the total cost for Model I (see table 6.3). This represents an increase of 1.5 percent in the total incremental cost of abatement and is in contrast to the 1.5 percent decrease in Model IX.

Input-Output Multipliers

The emphasis in Leontief (1970) is on the secondary rounds of demand for inputs, which result from the initial direct demand. Liu (1968, table V–1) has found, for the St. Louis economy, that a dollar increase in the output of the Chemical, Petroleum, and Rubber Products sector will result, after indirect demands are taken into account, in a $.024 increase in output of the Food, Tobacco, and Kindred Products sector, a $.014 increase in output of the Paper and Printing sector, a $1.061 increase in output of the Chemical, Petroleum, and Rubber Products sector itself, etc. Representing these multipliers by a $t \times t$ matrix **L** (which in this case is 23 × 23) whose typical element L_{ij} denotes the increase in output of sector i resulting from a dollar increase in output of sector j, the Linear Programming Model may be rewritten as follows:

Table 6.5

Breakdown of total incremental costs of abatement for the activity solution of Model I (millions of dollars)

Cost of inputs from the chemical, petroleum, and rubber products sector	.4
Cost of inputs from the machinery sector	6.7
Cost of inputs from the transportation equipment sector	8.7
Cost of inputs from the mining sector	− 12.2
Cost of inputs from the transportation, communications, and utilities sector	
Electric power only (296 million kilowatt hours)	3.2
Natural gas, etc.	12.3
Cost of inputs from the household sector	4.9
Cost of miscellaneous unallocated inputs	2.8
Credit for recovered by-products[a]	− 13.3
Opportunity cost of capital (incremental investment of $192 million)	19.2
Scarcity premium for natural gas (14,193 million cubic feet for air pollution control)	2.6
Total incremental cost of abatement	35.3

[a]This includes recovered sulfuric acid ($11.8 million), steam, coke-oven gas, leaf mulch, recovered sinter, fertilizers, gravel, etc.

Table 6.6

Tons of coal burned in the Meramec power plant (source number 41) per million dollars of sales by sectors of the St. Louis economy

Food, tobacco, and kindred products	7.93
Textiles and apparel	2.55
Lumber and furniture	4.95
Paper and printing	10.60
Chemicals, petroleum, and rubber products*	20.01
Lumber products	6.06
Stone, clay, and glass	55.06
Primary metals	38.58
Fabricated metals	9.18
Machinery (nonelectric)*	4.44
Electrical machinery	10.46
Transportation equipment*	6.23
Miscellaneous manufacturing	9.19
Agriculture	3.91
Mining*	17.46
Construction	3.66
Electric Power*	2,600.00
Wholesale trade services	8.03
Retail trade services	9.22
Finance, insurance and real estate	13.42
Business, personal, and other services	16.06
Household*	12.12
Local government	7.18

Note: Data from Kohn (1975a, pp. 252–254). In Model XI only those six sectors marked with an asterisk are used. (In Model XII all 23 rows of the **a** matrix are operative.)

Minimize **Cx**

subject to

$$[u - aLh]x = ay^0,$$
$$ex \leqslant \hat{f},$$
$$x \geqslant 0.$$

(6.22)

This version of the model, called Model XII, was implemented using the **L** matrix developed by Liu(1968, table V-1). When fully multiplied, the feedback effect of abatement increased total economic activity in the twenty-three sectors by $54 million. This augmented the pollution source levels and caused increases in the projected emission flows of the five pollutants, ranging from .2 to 2.8 percent of the allowable flows. As a consequence, the total cost of abatement for Model XII was $800,000 higher than the total cost with Model I (see table 6.3).[9]

Conclusions

In this chapter we have examined feedbacks of the control method solution on pollution source levels.[10] Neither of the feedback effects has a substantial impact on the total cost of abatement. Furthermore, the opposed feedbacks of increased output prices and derived demands for inputs are approximately equal and just about offset one another. By themselves, the individual feedbacks decrease or increase the incremental cost of abatement by three percent at the most (see table 6.3). Although the activity levels x_j^* in the solutions of Model IX and Model XII decreased or increased in differing proportions,[11] in no case was there a change in the basis matrix **B** of the optimal solution from Model I. Thus, the choice of abatement processes would not be affected by the corresponding feedbacks. It may be concluded that Models I and III are quite appropriate for policy making. The planning benefits obtained by modeling the feedback effects are probably not worth the additional costs of data collection and computation.

The models developed in this chapter place the Linear Programming Model in a general equilibrium context that is more realistic than that of the Pure Abatement model. The feedback of abatement costs prompts substitutions in consumption that are characteristic of a market economy, while the Leontief feedback interfaces the Linear Programming Model with the input-output version of general equilibrium analysis.

EFFICIENCY AND EQUITY IN AIR POLLUTION CONTROL

A government program for air pollution control is economically efficient if the air quality standards are optimal and if, given optimal outputs of other goods, they are achieved with a least-cost combination of control methods. This includes the costs of enforcement as well as the costs of abatement.

Economic efficiency also requires that selling prices be proportional to marginal social costs. The social costs include the external costs associated with the allowable emission flows.

There are various criteria of equity as applied to air pollution control. These relate to the distribution among households of the perceived benefits as well as the opportunity costs of abatement. There are also considerations of equity in the treatment of individual firms.

In the preceding chapters of this book, we have applied empirical analysis to the interpretation of optimal pollutant concentrations and cost minimization. Some observations have been made on the sensitivity of enforcement costs to pollution control regulations. Although the impact of abatement costs on final prices has been considered, it was not determined whether final price ratios were in fact proportional to marginal social costs. Finally, very little of empirical content has yet been said about equity.

In this chapter we examine enforcement costs, the development of pollution control technology, and relative price effects. These are considerations which, like the minimization of the total cost of abatement, are related to economic efficiency. We also examine some empirical data which relate to equity in pollution control. For purposes of illustration, we compare three alternative government programs for pollution control and rank them in terms of criteria for efficiency and equity. Finally, we combine equity and efficiency objectives in the structure of the Linear Programming Model.

Alternative Government Programs for Controlling Air Pollution

The primary objective of the regulatory agency is to reduce pollutant concentrations to optimal levels. There are alternative programs which may be used. These are (1) Least-Cost Standards, (2) Pigouvian Fees, and (3) Politically Feasible Standards. The first two of these programs are directly related to the Linear Programming Model. The solution x^* of that model defines a set of efficient production-abatement activity levels. These activities can be translated into a matrix of emission standards $\mathbf{E}$, where E_j^i denotes the legal allowable flow of pollutant i per unit of source activity j. If, for example, $\mathbf{x}^*$ includes the following:

$$x_{2b} = s_2,$$
$$x_{3a} = \gamma s_3, \qquad\qquad (7.1)$$
$$x_{3d} = (1 - \gamma)s_3,$$

where γ is some positive fraction, the corresponding emission standards for pollutant i would be

$$E_2^i = e_{2b}^i,$$
$$\qquad\qquad (7.2)$$
$$E_3^i = \gamma e_{3a}^i + (1 - \gamma)e_{3d}^i.$$

This includes the possibility of a zero emission rate, which de Nevers (1977, p. 198) calls a "prohibitive standard." The regulatory agency could control pollution by requiring firms and households to reduce emissions to the efficient rates. This program of pollution control is called *Least-Cost Standards.*

The solution of the same Linear Programming Model includes a set of pollutant shadow prices. These may serve as emission fees. Assuming that the levels of polluting activities are fixed, as in the Pure Abatement model, the imposition of emission fees equal to the shadow prices would prompt polluters to voluntarily[1] institute the control activities denoted by $\mathbf{x}^*$. Consider, for example, a sulfuric acid plant built prior to 1967 (see source 78 in table 2.1). For this source, abatement costs plus emission charges, given the base-year abatement process, would be

$$\$.16 + (45.0)(\$.02193)$$
$$+ (.4)(\$.07748) \cong \$1.18 \text{ per ton} \qquad\qquad (7.3)$$

of sulfuric acid produced. However, by installing the double contact process 78b these costs would decline to

$$\$.98 + (4.5)(\$.02193)$$
$$+ (.4)(\$.07748) \cong \$1.11. \qquad\qquad (7.4)$$

Accordingly, the sulfuric acid producer would minimize total costs by using process 78b. This program of pollution control is called *Pigouvian Fees* in recognition of the economist who first conceived of it (see Pigou, 1960, p. 381).

Both the Least-Cost Standards and the Pigouvian Fees programs are based on the same linear programming model.[2] In chapter 3 we interpret the significance of the shadow prices π^i and π_i and the reduced cost ρ_j in terms of least-cost considerations. The shadow prices also have an interpretation in the context of Pigouvian Fees. The summation $\sum_i^m \pi_i s_i$ is the total of efficient abatement outlays and emission fees, the summation $\sum_i^p \pi^i f^i$ is total fee revenue collected by the government, and ρ_j is the unit loss for implementing an inefficient process j. (It follows from the discussion above that the reduced cost for activity 78a is approximately \$.07.)

The third program of pollution control is representative of regulations actually imposed by government agencies. In general, these tend to require some degree of abatement by every pollution source. Although not equiproportional, neither are they least-cost. This third category of pollution control programs is called *Politically Feasible Standards*.[3] In our analysis, we shall take it for granted that the regulatory agency has no power to make lump sum transfers to compensate for inequities.

Direct Costs of Abatement

Least-Cost Standards and, in theory, Pigouvian Fees would achieve improved air quality at the least abatement cost. It was estimated in chapter 3 that the total incremental cost of abatement with a least-cost solution is 25 percent less than the cost incurred with Politically Feasible Standards. This is less than the 50 percent estimate of Atkinson and Lewis (1974), but somewhat higher than the 18 percent estimate of Tihansky (1973, p. 349). Clearly, the total direct cost of abatement is highest under Politically Feasible Standards.

It has been suggested that the total cost of abatement under Pigouvian Fees would be less than that under Least-Cost Standards. This would be the case because firms would be aware of more options for reducing pollution than a government agency could discover and mandate. Granted that this is true, the total cost of abatement would be least under a program of Pigouvian Fees. This ranking of the three programs is included in the first row of table 7.1.

Technological Development

The Linear Programming Model is based on the known technology of abatement. It is likely that the government program for controlling pollution

Table 7.1

Ranking of alternative programs for air pollution control in terms of efficiency and equity criteria

Criteria	Least-Cost Standards	Politically Feasible Standards	Pigouvian Fees
Direct costs of abatement	medium	highest total cost	lowest total cost[a]
Technological development	least favorable impact	medium	most favorable impact[a]
Regulatory cost	medium	lowest total cost[a]	highest total cost
Correction of relative price distortion	neutral	somewhat favorable	most favorable[a]
Equity consequences	unfavorable	unfavorable	neutral[a]

[a]Most favorable program for that criterion.

would have an impact on the long-run development of control technology. What is at issue here is the *long-run* minimization of the total cost of abatement. Zerbe (1970, p. 371), Schneider (1973), and Dewees (1974, p. 134) have argued that technological improvement is more likely to come about under Pigouvian Fees. This is the case because with fees there is an incentive to reduce emissions below levels that would be allowable under standards.

Alternatively, there is counter evidence that standards and regulations compel firms that might have preferred paying only fees to abate and thereby gain the kind of experience which promotes advances in technology.[4]

Granted that Zerbe, Schneider, and Dewees are correct, the Pigouvian Fee program is given the most favorable ranking in table 7.1 for promoting technological change. Granted also that experience precedes technological breakthroughs, it follows that Politically Feasible Standards, which require all polluting firms to abate, will result in more firms gaining experience in abatement than would Least-Cost Standards. Therefore the latter program is given the least favorable ranking in this category in table 7.1.

Regulatory Costs

The regulatory costs of administration and enforcement are not included in the parameters of the Linear Programming Model (see chapter 3). However, an effort was made, with the help of air pollution control officials in the St. Louis airshed, to compare the cost of administering Least-Cost Standards, based on the solution of Model I, with the cost of administering a corresponding programs of Pigouvian Fees.

The costs of the latter are much larger because of greater information and accounting requirements. In the case of Standards, not all pollution sources would necessarily be required to implement control measures, so that these sources need be checked infrequently if at all. For those sources required to reduce their emission rates, periodic checks for compliance may be sufficient. In the case of pollution control by Pigouvian Fees, all polluting sources must be continuously monitored, not only to determine their rates of emissions with respect to output, but to determine their output over time for computing total emission flows. In addition there are the transaction costs of periodic billing.

The estimated costs of administering Least-Cost Standards for stationary sources alone would have been \$500,000 in 1975 as compared to \$4 million for Pigouvian Fees. The latter is an understatement because it includes no costs for taxing emissions from coke ovens, incinerators, open burning, leaks in gas lines and refinery valves or evaporation from bulk storage tanks, as these would be almost impossible to monitor; nor does it include costs for taxing the emissions caused by residential furnaces.

The above estimates of administrative costs assume ready compliance with either program and therefore no litigation costs. However, the Least-Cost Standards would be considered inequitable to the extent that there are significant sources of pollution for which abatement is inefficient. Because these sources would not be required to abate under Least-Cost Standards this program would be considered unfair, and would be resisted by those who had to carry the burden.

This problem would not exist with Pigouvian Fees because the fee per unit of pollutant is the same for all firms and abatement is a voluntary decision. On the other hand, the legal costs associated with Pigouvian Fees would be high because of difficulties in measuring the emission flows on which the charges are based. According to Lillis and Schueneman (1975, p. 808) the *best* continuous monitoring devices are only accurate within a ± 20 percent range. Because a substantial amount of money is based on such emission flows, there would surely be extensive controversy and litigation.

By definition Politically Feasible Standards spread the burden of abatement so as to minimize legal resistance to pollution control. Although there are more sources to monitor than under Least-Cost Standards, it appears on balance that enforcement costs are lowest for Politically Feasible Standards. Because a complete monitoring of all emission flows is almost impossible, the enforcement costs are highest for Pigouvian Fees.[5] This ranking of the programs is included in table 7.1.

Correction of Relative Price Distortion

A condition of economic efficiency is that selling prices are proportional to marginal social costs. In the pure abatement model developed in chapter 1, this is achieved by either Least-Cost Standards or Pigouvian Fees. In the general equilibrium models introduced in chapter 6, individual goods are characterized by different ratios of emission flows to dollar costs. If selling prices are proportional to production costs and there is a government program for pollution control, then relative prices are distorted to the extent that the marginal social costs of pollution are not reflected in selling prices.

Each of the three programs for pollution control described in this chapter will effect relative prices differently. We shall evaluate whether the impact is one which reduces or increases the preregulatory distortion in relative price. That reductions in distortion are welfare-increasing has been demonstrated by Foster and Sonneschein (1970), Kawamata (1974), and Rader (1974).

Emission Standards and Relative Price

We first consider the impact of Least-Cost Standards, based on the solution of Model I, upon relative prices. There is no reason to expect that price ratios under Least-Cost Standards would be less distorted than preregulatory prices. It is even possible that they would be more distorted.

This analysis of relative price effects is based on the categorization of economic activities that was developed in the preceding chapter.

The total incremental cost of abatement in the solution of Model I is $35.3 million. It is borne by economic sectors as shown in table 7.2. This set of standards would increase the private production-abatement costs for these sectors by the amounts shown in column 2. The balance of the incremental total cost of abatement consists of the nonprivate costs in column 3; these would not be viewed as accounting costs by firms and households.

Fourteen of the categories in table 7.2 are sufficiently narrow to treat as single goods. These are included in table 7.3. The total base-year production

Table 7.2
Incremental costs of abatement in the least-cost solution of Model I (millions of dollars)

(1) Economic activities	(2) Private costs of abatement	(3) Nonprivate costs of abatement[a]
Consumption activities		
Automobile driving	16.5	0
Residential heating	n	1.4
Miscellaneous	n	0
Service activities		
Bus transportation	0	0
Air transportation	0	0
Dry cleaning	0	0
Combustion of residential refuse	.4	0
Commercial and institutional activities	.3	.1
Industrial activities		
Electric power generation	10.1	0
Oil refining	.5	n
Sulfuric acid production	1.0	n
All other chemical industries	1.1	.3
Cement production	.1	n
Steel production	.9	.1
Iron and steel foundries	.1	n
Lead smelting	n	n
Truck, barge, rail transportation	0	0
Miscellaneous industrial activities	1.7	.7
Totals	32.7	2.6

Note: n indicates less than $50,000. Data from Kohn (1977a).
[a]The nonprivate cost of abatement is the scarcity premium for natural gas used for pollution control.

abatement costs are the market values in column 2; the incremental private costs associated with the least-cost solution are the values in column 3; and the total private costs are the sums in column 5. We shall assume that the increased costs are reflected in higher prices and that the entries in column 5 represent market values for the same quantities of output that would have been produced in the absence of regulations. Finally, it is assumed that the total social cost associated with each good is the value in column 7. This is the sum of the preregulatory cost in column 2, the private and nonprivate costs of abatement in columns 3 and 4, and the monetary value of the disutility of pollution in column 6. The latter is the dot product of the allowable flows[6] shown in table 6.2 and the respective pollutant shadow prices in table 3.3. The rationale for representing the dollar value of the marginal disutility of each pollutant by its shadow price from the solution of the Linear Programming Model is explained in chapter 5.

In this analysis we assume that selling prices are proportional to the market values shown. This follows from the assumption that the values in each column of table 7.3 are based on the same preregulatory quantities of physical output. We shall denote the preregulatory selling prices of goods i and j by p_i, and p_j, the selling prices under Least-Cost Standards by p_i^s and p_j^s, and the prices which equal marginal social costs by p_i^* and p_j^*. If, for any pair of goods,

$$p_i/p_j < p_i^s/p_j^s \leqslant p_i^*/p_j^*, \tag{7.5}$$

if follows that emission standards reduce the distortion in the ijth price ratio.[7] If the price ratios are such that

$$p_i^s/p_j^s < p_i/p_j < p_i^*/p_j^*, \tag{7.6}$$

then the effect of standards is to increase price distortion. There is also the possibility that

$$p_i/p_j < p_i^*/p_j^* < p_i^s/p_j^s. \tag{7.7}$$

In this case, if

$$|p_i/p_j - p_i^*/p_j^*| > |p_i^s/p_j^s - p_i^*/p_j^*|, \tag{7.8}$$

then (7.7) implies a reduction in the distortion of relative price. If the inequality in (7.8) is reversed, then (7.7) implies an increase in the distortion.

In an economy with many goods a reduction in price distortion between

Table 7.3

Market value of economic activities with a least-cost set of emission standards and total social costs (millions of dollars)

(1) Economic activities	(2) Estimated market value of economic activities in 1975 in the absence of abate- ment	(3) Private annual cost of abate- ment with least-cost set of emission standards	(4) Non- private annual costs of abate- ment	(5) Market value of economic activities with least- cost set of emission standards	(6) Dollar value of the disutility of pollution	(7) Total social cost of economic activities including abatement and dis- utility costs
Automobile driving	1,940.0	16.5	0	1,956.5	45.9	2,002.4
Residential heating	304.5	n	1.4	304.5	9.9	315.8
Bus transporta-tion	30.5	0	0	30.5	.6	31.1
Air transporta-tion	279.2	0	0	279.2	1.7	280.9
Dry cleaning	12.6	0	0	12.6	.2	12.8
Electric power generation	419.6	10.1	0	429.7	33.3	463.0
Oil refining	660.5	.5	n	661.0	10.2	671.2
Sulfuric acid production	28.1	1.0	n	29.1	.4	29.5
All other chemicals	1,235.0	1.1	.3	1,236.1	9.9	1,246.3
Cement production	49.6	.1	n	49.7	2.0	51.7
Steel production	279.5	.9	.1	280.4	3.6	284.1
Iron and steel foundries	154.0	.1	n	154.1	1.1	155.2
Lead smelting	108.1	n	n	108.1	.8	108.9
Truck, rail and barge	393.2	0	0	393.2	6.0	399.2

Note: n indicates less than $50,000. Data from Kohn (1977a).

two goods may increase the price distortion between one of these goods and a third good. In the case of an economy with t goods, there are $(t/2)(t-1)$ price ratios; and the imposition of emission standards is likely to both increase and decrease price distortion.

In assessing the relative price effects of the least-cost set of emission standards for the St. Louis airshed, we shall assume that the ratio of preregulatory selling prices of any two goods, p_i/p_j, is proportional to the respective entries in column 2 of table 7.3; that the ratio of selling prices after standards are implemented, p_i^s/p_j^s, is proportional to the respective entries in column 5; and that the ratio of social costs, p_i^*/p_j^*, is proportional to the respective entries in column 7.

We shall value the preregulatory distortion of the ijth price ratio by

$$T_{ij} = p_i/p_j - p_i^*/p_j^* \tag{7.9}$$

and the change in that distortion by

$$t_{ij} = p_i^s/p_j^s - p_i/p_j. \tag{7.10}$$

There are 91 different price ratios in this model and, for consistency, the larger of each pair of prices is taken as the denominator. The proportional change in the ijth price ratio is

$$\theta_{ij} = t_{ij}/T_{ij}. \tag{7.11}$$

If θ_{ij} is negative, there is a reduction under Least-Cost Standards of the ijth price ratio. An example of this is the case of automobile driving and electric power generation. In this case, the sequence of ratios corresponding to (7.5) is

$$\frac{\$419.6}{\$1940} < \frac{\$429.7}{\$1956.5} < \frac{\$463.0}{\$2002.4}. \tag{7.12}$$

It follows that $T_{ij} = -.0149339$, $t_{ij} = .0033382$, and $\theta_{ij} = -.22$.

If the value of θ_{ij} is positive, this implies that there is an increase in distortion, as in (7.6) above. An example of this is the case of automobile driving and residential heating, where $T_{ij} = -.0007519$, $t_{ij} = -.0013237$, and $\theta_{ij} = +1.76$.

A sequence of inequalities such as (7.7) implies that relative price is shifted in the correct direction. This is distortion-reducing provided that (7.8) is satisfied. For cases in which (7.7) is applicable, the change in relative price distortion is

Table 7.4

Changes in relative price distortions

	Automobile driving	Residential heating	Bus transportation	Air transportation	Dry cleaning	Electric power generation
Automobile driving						
Residential heating	+1.76					
Bus transportation	−.70	0				
Air transportation	−.33	0	0			
Dry cleaning	−.53	0	0	0		
Electric power generation	−.22	−.39	−.31	−.27	−.30	
Oil refining	−.50	+.04	+.22	−.08	+.36	−.27
Sulfuric acid production	−.43	+.90	−.80	−.82	−.94	+.23
All other chemicals	−.34	+.03	+.09	−.29	+.13	−.25
Cement production	+.65	−.40	−.09	−.06	−.08	−.39
Steel production	−.34	+.16	+1.02	−.31	+3.57	−.25
Iron and steel foundries	−.33	+.02	+.06	−.38	+.08	−.26
Lead smelting	−.35	0	0	0	0	−.27
Truck, rail, and barge	−.51	0	0	0	0	−.29

Oil refining	Sulfuric acid production	All other chemicals	Cement production	Steel production	Iron and steel foundries	Lead smelting	Truck, rail, and barge
−.95							
+.01	−.86						
−.05	+2.67	−.03					
+7.69	−.98	−.32	+.05				
−.01	−.84	−.18	−.04	−.30			
−.09	−.85	−.51	−.06	−.36	−.33		
−.82	−.95	+.15	−.08	+.73	+.09	0	

$$t_{ij} = 2(p_i^*/p_j^*) - p_i/p_j - p_i^s/p_j^s. \tag{7.13}$$

An example of (7.8) is sulfuric acid production and oil refining. In this case, $T_{ij} = -.0014076$, $t_{ij} = .0013345$, and $\theta_{ij} = -.95$. An example in which the inequality in (7.8) is reversed occurs with steel production and dry cleaning. In that case, $T_{ij} = .000026$, $t_{ij} = .0000927$, and $\theta_{ij} = +3.57$.

For those sectors where the price ratios under Least-cost Standards are not significantly different from the preregulatory price ratios, the t_{ij} and therefore the θ_{ij} are equal to zero. Table 7.4 contains the values of θ_{ij} for the fourteen economic categories in this study.

There are 76 nonzero entries in table 7.4, of which 53 are negative (distortion-reducing) and 23 positive (distortion-increasing). To obtain an overall measure μ of the impact of this set of emission standards on relative price distortion, the θ_{ij} are weighted, summed, and normalized as follows:

$$\mu = \frac{\sum\limits_{i}^{13} \sum\limits_{j=i+1}^{14} \theta_{ij}(p_i y_i)(p_j y_j)}{\sum\limits_{i}^{13} \sum\limits_{j=i+1}^{14} (p_i y_i)(p_j y_j)}. \tag{7.14}$$

To reflect the relative significance of the various price ratios in consumption decisions, each θ_{ij} in (7.14) is weighted by the product of the preregulatory market values of goods i and j. These are the values in column 2 of table 7.3.

The value of μ obtained with the above formula is

$$\mu \cong \frac{-980 \times 10^{-15}}{14,200 \times 10^{-15}} \cong -.07. \tag{7.15}$$

The maximum possible value of μ would be -1.0, which would result if each θ_{ij} were equal to -1.0. Because μ is approximately zero, we may tentatively conclude that, on balance, the effect of these emission standards on relative price ratios is essentially neutral. Although this set of standards would achieve the air quality goals at least cost, it would not alter relative prices in such a way as to reflect the marginal damage associated with allowable emission flows. Although this numerical test is a crude one, we shall assume that the result is generally correct.

There is no numerical evidence to indicate that the impact on relative price distortion would be different for Politically Feasible Standards. However, it is reasonable to assume that the costs of abatement, imposed by Politically Feasible Standards, have some positive relationship to the impact of the allowable emission flows from the individual sources. This would support the notion of their being "fair." Accordingly, we shall assume that Politically Feasible Standards have a more favorable effect than do Least-Cost Standards in reducing the preregulatory distortions in relative price ratios.

Pigouvian Fees and Relative Price

If we assume that some omniscient regulatory agency (the analogue of Samuelson's omniscient planner, 1969, pp. 102–104) knew all of the parameters of the economy, including each household's utility function and each firm's production function, and could determine in advance the Pareto optimal levels of output y^*, the optimal pollutant flows f^*, and, using a version of the Linear Programming Model, the corresponding pollutant shadow prices π, this agency could, at the outset, establish the correct Pigouvian fees. If the appropriate model were indeed Model I, the Pigouvian fees would be

$$\phi^c = \$ .00428 \text{ per pound of carbon monoxide,}$$
$$\phi^h = \$ .02476 \text{ per pound of hydrocarbons,}$$
$$\phi^n = \$ .32639 \text{ per pound of nitrogen oxides,} \qquad (7.16)$$
$$\phi^s = \$ .02193 \text{ per pound of sulfur dioxide,}$$
$$\phi^p = \$ .07748 \text{ per pound of particulates.}$$

The imposition of these fees would increase private costs by the quantities in column 6 of table 7.3. In the absence of nonprivate costs, the ratio of the ijth pair of market prices would be identical to the ratio of marginal social costs p_i^*/p_j^*. Accordingly, the value of μ, which measures the reduction in preregulatory price distortion, would be -1.0.

If the regulatory agency accounted for the Leontief feedback, the shadow prices would be those of Model XII, or

$$\phi^c = \$ .00432 \text{ per pound of carbon monoxide,}$$
$$\phi^h = \$ .02482 \text{ per pound of hydrocarbons,}$$
$$\phi^n = \$ .33333 \text{ per pound of nitrogen oxides,} \qquad (7.17)$$
$$\phi^s = \$ .02220 \text{ per pound of sulfur dioxide,}$$
$$\phi^p = \$ .07941 \text{ per pound of particulates.}$$

These fees are higher than those in (7.16) because the imput requirements **for abatement increase the flow of emissions and hence the required abate**ment. As a consequence of these emission fees, the selling price of the output of the Transportation Equipment sector would increase **by .5 percent; the** selling price of the output of the Chemical, Petroleum, and Rubber Products sector would increase by 1.8 percent; and the price of electricity would increase by 17.1 percent (see Kohn 1975a, pp. 265–266). This is the type of price corrective envisaged by Pigou (1960, p. 381).

In theory, then, the relative price effect of Pigouvian Fees is very favorable. The ranking of the three governmental programs in terms of their corrective effect on relative price distortion is contained in table 7.1. It should be noted that this evaluation is based on the assumption of perfectly competitive markets and would not necessarily be the same for an economy in which polluting firms are monopolies.

Pigouvian Fees and Land Use

Tietenberg (1974) has argued that pollution fees should play a role in the allocation of urban land. Given a diffusion formula such as (4.14), the contribution of a pound of emissions at distance k from the receptor station to the concentration of that pollutant at the receptor (where $k = 0$) increases geometrically as distance decreases. Assuming that the adverse impact of pollution is proportional to the measured concentration, it follows that the emission fee rate for any source should increase as distance decreases. As a result, the rent which a firm could bid for any particular location would be appropriately diminished by the marginal external costs of its emissions.[8] Accordingly, Pigouvian Fees would promote the efficient allocation of urban land among competing uses.

This concept was tested using Linear Programming Model V. The location of Granite City Steel Company was varied along a ray northeast of the CAMP station. The corresponding fee charges for particulates at selected locations, assuming that the air quality standards are achieved at least-cost, are shown in table 7.5. The diffusion factors in column 2 increase because the closer a sorce is to the CAMP station, the greater is the contribution of one pound of particulate emissions to the concentration at the receptor. Granite City Steel Company emits a large quantity of pollution; accordingly, the particulate shadow price itself increases (see column 3) as its distance to the source decreases.[9] The Pigouvian fee in column 4 is the product of columns 2 and 3.

The total fees, per acre of land occupied, that would be paid by the steel company at each location are shown in table 7.6. At its present location 6 miles from the CAMP station the fees would total $1,457 per acre. This is

Table 7.5
Pigouvian fee for particulates at selected distances northeast of the CAMP station

(1)	(2)	(3)	(4)
Distance (miles)	Calibrated diffusion factor for particulates[a] (μg/m^3 per pound)	Shadow price for the particulate requirement in Model V[b] (dollar per μg/m^3)	Pigouvian fee for particulates (dollar per pound)
10	476×10^{-10}	583,000	.02775
8	729×10^{-10}	588,000	.04287
6	$1,259 \times 10^{-10}$	593,000	.07466
4	$2,699 \times 10^{-10}$	779,000	.21025
2	$9,553 \times 10^{-10}$	912,000	.87123

[a]This is equivalent to the factor m_j times the corrective constant 1.86 (see note 12 in chapter 4).
[b]The shadow price is based on the assumption that Granite City Steel Company is located at the corresponding distance.

Table 7.6
Total pollution fees that would be paid by Granite City Steel Company at various distances northeast of the CAMP station

Distance (miles)	Total fees per acre of land occupied (dollars)
10	660
8	1,029
6	1,457
4	3,109
2	19,581

Note: See Kohn (1974a). This includes Pigouvian fees on all of the pollutants emitted.

surprisingly high when compared to the value of the land itself, which is estimated to be $10,000 per acre. In this case, the pollution fees probably exceed the imputed annual rental value of the land. This illustrates that unpriced external costs of land-use may be more substantial than is generally believed.

A problem with Model V is that there is a single receptor and hence a single measure of air quality. A better model for this purpose is one in which there is a receptor in each residential area, as in the Guldmann model (see table 4.7). This example does illustrate a problem with Pigouvian Fees which vary by location. Ferrar et al. (1975) suggest that there would be a great deal of opposition to a system in which competing firms paid different fees for the same pollutant, solely because they were in different locations of the same airshed. As we have noted earlier, the appearance of inequity raises opposition to the government's program for pollution control and boosts the regulatory costs. This reaffirms our adverse ranking in table 7.1 for the regulatory costs of Pigouvian Fees.

The Meaning of Equity

An allocation of inputs between firms and outputs among households is economically efficient if there is no alternative allocation which would increase the utility of some household without reducing the utility of some other household. This concept of economic efficiency is extended in chapter 1 to include an optimal level of air pollution. An economically efficient allocation of resources exists if the combination of attained utility levels is on the utility possibility frontier Such a frontier, for a two-household economy, is illustrated in figure 7.1.

Prior to abatement, the set of utilities is represented by the point P. In the welfare economic models in chapter 1, we maximize U^1 for a given value of U^2. Assuming that the given value is $\bar{U}^2$ in figure 7.1, the optimal solution results in the combination R. Presumably such a solution would be achieved in a competitive market economy with appropriate Pigouvian fees and lump sum transfers of the fee revenue to the households. However, it is unlikely that governments could effect the compensating lump sum transfers,[10] certainly not at the local airshed level at which regulations are enacted. Thus, a government program of pollution control might result in a move from point P (in figure 7.1) to one such as S, in which one of the households is made worse off.

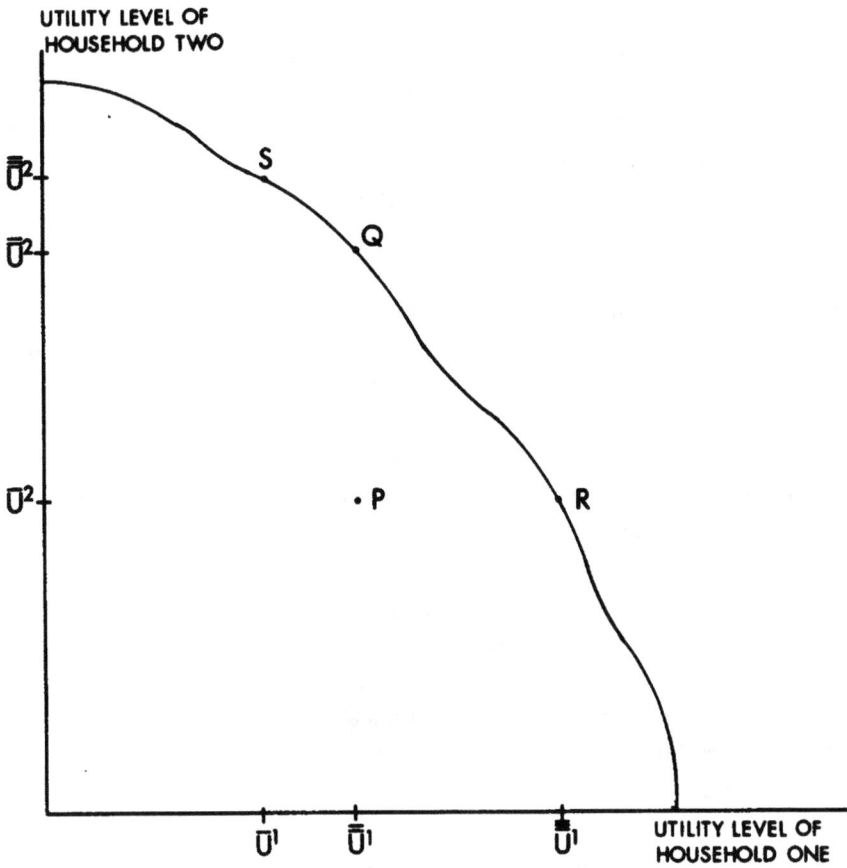

Figure 7.1
Utility possibility frontier

According to the benefit theory of equity (see Kohn, 1975b, p. 133), we would prefer that the move from P was to some point on the frontier between Q and R. This would be the case if the reductions in real income (in terms of consumption of private goods) are borne by the individual households in proportion to their demands for cleaner air.

While this definition of equity is useful, it has limitations. One of the households may live on the perimeter, or outside of, the airshed, where there is little or no perceived improvement in air quality. Nevertheless, such a household's consumption of pollution-related goods indirectly contributes to the contamination of air for other households, and from another perspective (see Kohn, 1975b, p. 129) we would view this as a prior inequity that should be corrected. Then the combination S (assuming that the outlying household is the one made worse off) would be equitable as well as efficient, a value judgment that would be reinforced if the poor air quality were in the residential areas of low income households.[11] In our desire to prevent any reduction of the buying power of these households, we may prefer that the higher income households pay more for air quality than the benefits are worth to them. There is a tradition in welfare economics, starting with Kaldor (1939, p. 550), that a movement from a point such as P in figure 7.1 to one such as S is consistent with Pareto optimality; although one household is made worse off, and remains so, what counts is that the potential has been achieved for both to be better off.

It is clear from the above that our interpretation of equity in pollution control will depend on what the particular inequity is and who is being made worse off.

Empirical Research on the Equity of Standards

The implementation of emission standards increases production costs and, consequently, market prices. This reduces the purchasing power of households, particularly those households who have a greater preference for goods whose relative prices are increased. The gross real income effect has been examined by Dorfman (1976) and the relative price effect by Yan et al. (1975). Their results will be examined here.

Dorfman (1976) has evaluated the equity of the current national pollution control program. He has compared the resulting increases in the cost of living for various income classes with the quantity of money which households in each income class would have voluntarily paid for the corresponding improvement in environmental quality, and has concluded that

the cost of the pollution control program to middle bracket families was just about what they would be willing to pay to obtain a clean environment. Lower bracket families, on the average, were required to pay some $60 more per family than they regarded environmental cleanup as worth, while the average burden on the upper income families was about $60 less than they said they would be willing to contribute to obtain a clean environment.

Whereas Dorfman focused on lump sum increases in the cost of living, Yan et al. (1975) consider the increases in price of individual goods and the relative importance of these goods in the family budget of each income class. Yan found that the goods whose selling price increased most as a consequence of air pollution control represented a larger proportion of total consumption expenditures for middle income than for lower or upper income groups. The increased cost of living for middle income groups approached 3 percent, as compared to 2 percent for the highest and lowest income groups in their study.[12] However, this research did not consider the perceived value of a cleaner environment to the different income groups, as did Dorfman's.

According to Dorfman, pollution abatement results in an allocation corresponding to S in figure 7.1, and it is the lower income household that is worse off. It appears that Least-Cost Standards and Politically Feasible Standards are both regressive. As a result of this, and because the demand for air quality is income-elastic, both programs are given an unfavorable ranking for equity in table 7.1.

Equity of Pigouvian Fees

We have seen in chapter 1 that the fee revenue, paid to the government, must be transferred to households if they are to have sufficient purchasing power to buy the entire output of the producing sector. In theory, these lump sum transfers could be distributed in such a way as to correct for any inequities in the price and income effects of the pollution control program relative to the perceived benefits. Thus, a household which has relatively little interest in clean air but is penalized by higher prices, particularly of goods for which it had relatively high favor, could receive a larger share of the fee revenue.

The revenue from Pigouvian Fees could be substantial. In terms of Model I, it would be the dot product of the allowable emission flows $\hat{f}^i$ and the corresponding pollutant shadow prices π^i, a sum in excess of \$150 million a year.[13] This is three times the total cost of abatement for 1975, based on the solution of Model I. In reality, it is most likely that the revenue from pollution fees would not be distributed directly to households but would replace other forms of taxation.[14] If, for example, Pigouvian Fees could lessen dependence on the property and sales tax, which are themselves regressive (see Mills, 1972, p. 154), the equity effect would be at least neutral. Thus, Pigouvian Fees have the more favorable ranking with respect to equity in table 7.1.

Incorporating Equity Objectives in the Linear Programming Model

Pollution control may alter the earning capacities of households. For example, any workers who become unemployed as a consequence of pollution control most likely suffer a diminution in the present value of their labor resource. Model XIII, an extension of Model IX, was used to estimate the loss of jobs associated with the least-cost solution. Two types of job loss were identified. The first, represented by the coefficient w_j, denotes the loss in employment associated with one unit of abatement activity j. For example, $w_{17e} = .0002$ jobs. This indicates that for each ton of coal (in source 17) replaced by natural gas, there is a loss of .0002 jobs (in nearby strip mines, in coal shipping and distribution, and in in-plant coal storage and firing operations). Because the analysis is concerned with the employment impact of alternative abatement activities, the w_j corresponding to base-year activities are zero. The coefficient, w_j is an element of a $1 \times n$ row vector $\mathbf{w}$.

The second type of job loss is a consequence of the feedback effect in Model IX, through higher prices, on the final output of goods. The declines in output are represented by the matrix product $\mathbf{vx}$, which was defined in the preceeding chapter. The consequent loss of jobs is $\mathbf{Wvx}$, where $\mathbf{W}$ is a $1 \times t$ row vector whose element W^k is the number of employees in the St. Louis region per unit of output of sector k. For example, the coefficient W^k for the cement industry is .00009 workers per barrel of cement produced; for the sulfuric acid industry, it is .00024 workers per ton of acid produced, etc. The total job displacement, D, is therefore

$$(\mathbf{w} + \mathbf{Wv})\mathbf{x} = D, \qquad (7.18)$$

where $\mathbf{x}$ is the solution of Model IX. The value of D thus derived was a loss of 735 jobs in the St. Louis region. Specifically,

$$wx^* = 155,$$
$$Wvx^* = 580. \tag{7.19}$$

The model does not include the possibility that some firms might dismiss their workers and move to other cities.[15] It was simply assumed that all firms would remain but that those affected by pollution control regulations would raise prices accordingly and experience some declines in quantities demanded.

In the aggregate, 735 jobs are relatively insignificant in a region in which employment exceeds one million persons. Furthermore, it should be noted that the alternative feedback model, XI, incorporating the direct demand for additional inputs, indicated that the labor requirements for abatement totaled 640 jobs.[16] Thus, the additional employment is approximately equal to the job loss. Nevertheless, for workers who lose their jobs there is a temporary loss of income, as well as a psychic cost. It is conceivable that a regulatory agency would be sensitive to this kind of disruption and would deviate from a least-cost set of emission standards in order to minimize job loss.

A compromise between efficiency and equity objectives was simulated with the Linear Programming Model. A solution was sought that would be almost as efficient as the least-cost solution but would minimize employment loss. Model IX was reformulated with total cost of abatement as a constraint and (7.18) serving as the new objective function. The revised model, called Model XIII, was used to find a solution x^D which achieves the air quality standards, within one percent of the least total incremental cost, with minimum job loss.[17] The total job loss in this model was $D = 690$. In exchange for an increase of approximately \$350,000 in the total annual incremental cost of abatement, 45 jobs were preserved.[18] If a tolerance of two percent (or \$700,000) in control costs were allowed, the total job displacement could be decreased to 680. Clearly, the opportunity cost of reducing job displacement rises.

It is conceivable that policy makers have implicit trade-offs between efficiency and equity objectives and that these influence their regulatory decisions. However, the example of Model XIII would discourage this. The cost of preserving 45 jobs is \$350,000 a year. But this is a built-in cost, which would be reflected year after year in excessive outlays for abatement. Were these jobs not protected, the disruption would have been temporary, and these same people eventually reemployed.

A stronger case can be made for treating individual firms equitably.[19] This consideration is integrated into the air pollution control model of Seinfeld and Kyan (1971, p. 178) with constraints which specify a "maximum daily cost of control to be imposed on (each) source." That polluting firms should be treated equitably has, as we already noted, an efficiency aspect. If a set of standards were considered unfair, this could cause firms to resist compliance, thereby increasing administrative and enforcement costs (see Downing and Watson, 1974).

One way in which the Linear Programming Model can be programmed to achieve equity between firms is to aggregate all sources which must be treated equally. The Siegel, Ehrenfeld, and Morganstern model (see table 3.2) goes still further in this direction by interlocking clusters of compatible control methods. An inescapable conclusion is that equity and efficiency are related. When administration and enforcement costs are accounted for, which they are not in the Linear Programming Model (see chapter 3), it is possible that a control solution, which is somewhat more costly but more equitable, is in fact the true least-cost solution.

Conclusions

Three alternative government programs for pollution control have been compared with respect to five criteria of efficiency and equity. In four of these categories, Pigouvian Fees are the most favorable. It is no wonder that so many economists have urged the adoption of the Pigouvian approach to pollution control.

When a total perspective is taken of the costs of abatement, it is no longer clear that Least-Cost Standards are, in fact, least cost. The regulatory costs of administration and, particularly, enforcement may be so high that Politically Feasible Standards are the best alternative. It may be that the high regulatory costs for Pigouvian Fees offset its superiority in the remaining four categories. This, despite the fact that the low direct costs of abatement, the corrective effect on relative price, and the distributional neutrality of Pigouvian Fees make this program so appealing.

This chapter suggests that the Linear Programming Model can be revised so that similar sources of pollution would be treated equally. In this way the optimal solution becomes a least-cost, politically feasible solution.

8 A FINAL APPRAISAL

Three themes are interwoven in this book, and it is useful to distinguish them. Each in fact represents a successive stage in the author's research interests. The first theme is *planning*, the second is *simulation*, and the third is *economic efficiency*. Each has policy implications, which are discussed in this chapter.

Airshed Planning

The original purpose of the Linear Programming Model was that it would be a tool for planning air pollution control strategy at the airshed level. This would be a simple model, comparable to Model I and based on the Larsen formula, that could be readily adapted from one airshed to another. Because of the interdependencies in abatement between the individual pollutants, linear programming would be especially useful.

The information requirements are not large. Data on the technology of abatement is published by the United States Environmental Protection Agency, and emission inventories, by source, are maintained by regulatory agencies. Implementing the model would be relatively easy.

Nevertheless, there has been little interest by regulatory agencies in this kind of model. This conclusion is articulated by Russell and Spofford (1977, p. 91) who find it "to be clear that our efficiency models were far less useful in actual decision making in government at all levels than the enthusiasm of practitioners would indicate."

The lack of interest in a linear programming model for air pollution control is unfortunate. Such a model could be useful for planning regulatory strategy at the airshed level. Of course, such a model assumes that pollution control within an airshed is the responsibility of a single agency, whereas this is often not the case. However, such a model could be a cooperative project of the separate agencies in the same airshed. Clearly, pollution does not stop at political boundaries, and it is economically efficient to have unified air quality management of an airshed.

The Linear Programming Model can also be useful to national agencies who are responsible for the control of automobile emissions. With this model, they can test the efficiency of the strategies which they sponsor. These include inspection-maintenance and traffic plans as well as automotive devices.

Perhaps the hesitation of government agencies to use a model such as that described in this book represents, in part, a rejection of the Larsen formula.[1] Yet this same formula is implicit in certain pollution control policies. For example, the current strategy for meeting the standards for automotive pollutants is based on proportional reductions of total emissions. A second example is the "offset rule," in which new sources cannot move into certain airsheds unless there is first an equal reduction of emission flows by existing sources. Both of these examples are consistent with a planning model based on total annual emission flows and the Larsen formula.

Russell and Spofford (1977, p. 91) explain the disinterest by governments in efficiency models by "the lack of distributional considerations which are important to all public decision makers, especially where public goods are involved." It is a limitation of the Linear Programming Model that the least-cost solution might appear inequitable to some sources and therefore would be difficult to translate into regulatory decisions. However, as we have already noted, this is a problem that can be resolved by redefining pollution sources and abatement activities within the structure of the Model.

Decisions will still be made in a political context, but there is no good reason why government agencies should not use efficiency models to enlighten the process. The models are far from perfect. Yet they can be used to resolve some difficult and controversial issues of pollution control policy. The examples given in chapter 3 are evidence of this.[2]

Simulation Model

The format and some major results of the Linear Programming Model were first published in 1970. Additional information on abatement technology and emission sources were becoming available and it would have been comparatively easy to revise the data. However, I decided that such revisions could be left to other investigators, and that the model should now be used for simulation purposes. There were many questions being asked about the economic consequences of air pollution control that could be answered with the Linear Programming Model. By retaining the original data, the solutions of the current and subsequent models could be interrelated. Furthermore, the emphasis would be on qualitative results and relative changes in numerical values. Nevertheless, this book has provided an opportunity to compare numerical results obtained here with those reported by other investigators, both for St. Louis and other regions. This ex post numerical verification indicates a substantial robustness for the Linear Programming Model.

A number of significant results are obtained using the Linear Programming Model for simulation. It is found that abatement of air pollution increases the flow of solid waste by a substantial amount. When the Model is programmed to account for joint-wastes, the effect on the optimal activity solution is quantitatively small; however, the new activity solution includes more emphasis on control methods which recycle recovered wastes.

The least-cost solution is compared to the regulatory solution for the St. Louis area. This simulation test suggests that the maximum saving in total costs of abatement obtainable with the least-cost solution is approximately 20 to 25 percent below that of the regulatory solution. However, this does not account for regulatory costs which would probably be higher for the least-cost solution. This again suggests the possibility of aggregating pollution sources in such a way as to achieve a somewhat higher "least-cost" solution with lower regulatory costs.

A stochastic element is incorporated in the Linear Programming Model. The cost of greater certainty that the air quality standards will be achieved is substantial. This provides a basis for the economic case for episode control strategies.

Alternative locations of a large source of pollution are simulated and two major results are obtained. Significant savings in the total cost of technological abatement are made possible by the strategic location of large emitters. Also, external costs associated with pollutant emissions may be substantial in comparison to the rental value of land.

Pollution source levels in the St. Louis airshed are projected to the year 1985. The total cost of reducing emission flows to the same allowable flows used in Model I are very large. However, there is some justification for allowing larger flows in 1985 if the size of the airshed increases (i.e., pollution sources are dispersed over a larger geographic area).

The interface of benefits and costs is simulated with the Linear Programming Model, using various pollution index systems. One tentative result is that for the given abatement budget, there should be more stringent standards for sulfur dioxide and particulates and less stringent standards for the remaining, essentially automotive, pollutants. This research discloses a potential nonconvexity which could pose a problem for benefit-effectiveness analysis.

The feedback of the costs of efficient abatement on pollution source levels is simulated, and there is a small decrease in the total cost of abatement.

A Final Appraisal

Table 8.1

Principal applications of the alternative versions of the Linear Programming Model

(1) Model number	(2) Brief description	(3) Airshed planning
I	The basic model with constraints on total emission flows	Especially useful for resolving issues of cost-effectiveness when there are interdependencies between pollutants
II	The basic model with penalties on joint-wastes	Illustrates the desirability of control processes that recycle recovered wastes
III	The basic model with constraints on ambient air concentrations	Incorporates the legal standards for the various pollutants
IV	The Stochastic model	
V	Calibrated Diffusion formula	Facilitates the development of a locational strategy for pollution control
VI	The 1985 Model	
VII	The Benefit-Cost model	
VIII	Benefit-Effectiveness model	Useful for evaluating the economic efficiency of pollutant standards
IX	Feedback of higher costs on demand for goods	
X	Feedback of abatement costs on fuel substitutions	

(4) Simulation modeling	(5) Economic efficiency analysis
	Relates to a general equilibrium model in which pollution originates in the production of intermediate goods that are required as inputs to production in fixed proportions to the labor input
Evaluates the relative significance of joint flows of solid, liquid, and thermal wastes	
Gives a reasonable estimate, from the cost side, of the lower limit for the benefits of pollution abatement	
Illustrates the high cost of greater certainty and the logic for episode control	
Shows that unpriced externalities may be large in comparison to land rental	Used to illustrate the resource allocative effect of Pigouvian Fees
Demonstrates the need for periodically updating the Larsen constants	
Provides results which are consistent with the assumption that the demand for environmental quality is income elastic	Demonstrates the relationship of optimal air quality standards to costs as well as to benefits
	Reveals a nonconvexity in the interface of abatement technology and nonlinear damage functions
Illustrates that this feedback effect is relatively small	Represents an attempt to model efficiency in a more realistic general equilibrium context
Illustrates the voluntary substitutions in production that are a response to pollution control regulations	

Table 8.1 (Continued)

(1) Model number	(2) Brief description	(3) Airshed planning
XI	Feedback of the primary demand for inputs for abatement	
XII	As Model XI with indirect demands	
XIII	Job-loss minimization	

However, there is an increase in total cost when the feedback of derived demands for inputs is simulated. It is concluded that these feedbacks are offsetting and need not be included in planning models.

The impact of a least-cost set of emission standards on relative price ratios is examined. It is concluded that any reduction in preregulatory relative price distortion is insignificant.

The job displacement caused by a least-cost set of emission standards is simulated. The observed trade-off of abatement cost for employment confirms the view that the primary objective of pollution control policy should be economic efficiency.

A program of Pigouvian Fees is examined. The gross annual fees collected by the government could be three times the total annual cost of abatement. This source of revenue could supplant other less desirable sources of revenue for local governments.

The above results are based on the original data used in the Linear Programming Model. Presumably, many of the same conclusions would be obtained with more current data. With the growing awareness of interdependencies between pollution control and economic variables, however, there will be a continuing need for empirical simulation models to answer new questions that will arise.

Economic Efficiency

In this book, a theory is developed which makes the Linear Programming

(4) Simulation modeling	(5) Economic efficiency analysis
Suggests that the Leontief impact is relatively small	
Indicates that the secondary impact of derived demands in less than the primary impact	
Illustrates the high cost of incorporating distributional equity consideration in the objective function	Demonstrates the trade-off between efficiency and equity objectives

Model a special case of a general equilibrium model. The conditions for economic efficiency apply to the Linear Programming Model, and the least-cost solution and pollutant shadow prices acquire additional significance.

The theory of optimal concentrations is based on welfare economics and incorporates a summation of the demand for air quality by individual households. The model does have theoretical limitations, for it is not clear that individuals have sufficient information to place a proper value on air quality. Nevertheless, the model of economic efficiency is useful and has policy implications. For example, it illustrates the desirability of a combined index of pollutants, as compared to the individual pollutant index systems which are being developed by the federal government.[3] The welfare economic model also illustrates the significant relationship between the optimal level of pollution and the distribution of income.

The most important emphasis of the theory is that optimal pollutant concentrations are a function of the costs of abatement as well as the benefits. If this principle were more widely recognized on all sides, there would be less controversy in the regulatory arena. It also follows that decisions on the level of environmental quality should take place at the local rather than the national level.[4] It is also consistent with economic efficiency that different airsheds would have different air quality standards.

In this final chapter, we have distinguished three major themes of this book. Each of the models presented in the preceding chapters are related to one or more of these themes. This is illustrated in table 8.1. This table provides a final review of some of the major results in the book and demonstrates the wide scope of the Linear Programming Model. In their entirety, the three themes indicate that the major emphasis of this book is not on solving the problem of air pollution in St. Louis, but rather on illustrating the usefulness of the linear programming framework for evaluating general impacts of alternative strategies and for relating these strategies to principles of economic efficiency.

NOTES

Notes to Chapter 1

1. The cardinal value of $U^i(\cdot)$ has no significance. All that matters is that one combination of outputs yields *more* utility to the ith household than another combination of outputs.

2. The reader may confirm that the cost of reducing the pollution level, q, by one unit, holding y_2 constant, is $(c_{1c} - c_{1b})/(e_{1b} - e_{1c})$, by solving the following differential equations for ΔR:

$$\Delta x_{1b} + \Delta x_{1c} = \Delta y_1 = 0,$$
$$e_{1b}\Delta x_{1b} + e_{1c}\Delta x_{1c} = \Delta q = -1,$$
$$c_{1b}\Delta x_{1b} + c_{1c}\Delta x_{1c} = \Delta R - c_2\Delta y_2 = \Delta R.$$

We shall subsequently illustrate, with geometry, that no more than two processes for making good-one would be used simultaneously. The algebra of production-abatement, as expressed in linear activity analysis, is contained in Kohn (1975b, chapters 3, 4). This chapter draws extensively on chapters 1, 2, 3, 4, and 5 of that book.

3. We assume in this chapter that the transactions costs of government regulation are zero.

4. For there to be final outputs in this economy, it must be the case that $\alpha_j c_j < 1$ for any process j used to make the intermediate good.

5. This is demonstrated in the appendix to this chapter.

6. An increase in the allowable pollution level decreases the total cost of abatement.

7. These emission fees may be calculated from the shadow prices, $\pi^1 = 1/60$ and $\pi^2 = 1/30$, from model (1.57) by using a multiplier comparable to Ω above but adapted to the two-pollutant case.

Notes to Chapter 2

1. In the preceding chapter, the e_j^i represented contributions to a pollutant concentration q^i. Henceforth, the emission factors denote pounds of emission flows. The relationship between emission flows and pollutant concentrations is discussed in chapter 4.

2. The numbering of the activities in table 2.1 is different from that in Kohn (1969a). However, the ordering is essentially unchanged. To simplify table 2.1, certain of the control methods that were not efficient in Model I have been excluded. In fact, some of the excluded activities are efficient in subsequent extensions of the model. (For example, a dry collector, which is less efficient than a wet scrubber for controlling air pollution from a particular source, becomes more efficient when joint flows of water pollution are constrained.)

3. The data on carbon monoxide waste heat boilers were obtained from Danielson (1967, pp. 647, 650), Lunche et al. (1966, p.27), and from private communications with E. J. Sullivan of American Oil Co., George E. Sample of Shell Oil Co., and H. A. Lutz of Mobil Oil Corp. See Kohn (1969a, pp. 377, 8).

4. The sequence of marginal costs of particulate abatement, using consecutive pairs of processes, 86a, 86c, 86d, 86e, and 86f, is (after rounding): .01 < .02 < .06 < .17. The fact that 86b is excluded indicates that it is technically inefficient and should have been deleted

from table 2.1. It is a limitation of the model that the step-wise linear technology is not always a good approximation of abatement technology. In the case of automobile pollutants, Dewees (1974, pp. 59, 107) works with a nonlinear function.

5. This is based on emission factors in Venezia and Ozolins (1966, p. 471).

6. The Linear Programming Model can be expressed in an alternative format in which required *reductions* in total emission flows replace the allowable flows. In this format, which is used in Kohn (1969a), the emission factors such as e^i_{ja}, e^i_{jb}, and e^i_{jc} are replaced by abatement factors such as $(e^i_{jb} - e^i_{ja})$ and $(e^i_{jc} - e^i_{ja})$. The **x** vector does not include the base-year activity variables such as x_{1a}, x_{2a}, etc., but only the alternative activity variables. Consequently, the **Ux** = **s** constraint is replaced by **Ux** ≤ **s**. The format of the model based on allowable flows was found to be more convenient computationally.

7. It was assumed, for example, that the pollution from National Lead Company in south St. Louis arose entirely from the production of sulfuric acid and the combustion of coal. Now it is known that there are significant emissions from the grinding of titanium oxide ore and from the digestors in which the ore and sulfuric acid are mixed.

8. This example was developed in Kohn (1969b) and was subsequently used by Cooper and Steinberg (1974, pp. 37–38).

9. The reader may solve this linear programming problem with the Simplex method (see Hadley, 1962, pp. 124–130), or by solving the two constraint equations simultaneously for each pair of activity variables. The inequality is converted to an equation by adding a slack variable (see Hadley, 1962, pp. 72–76). The optimal solution is one of the basic feasible solutions.

Notes to Chapter 3

1. A set of punched input cards containing the data in table 2.1 and (2.12) is available from the author at a nominal charge. These together with the appropriate MPS/360 program cards will generate the output for Model I discussed in this chapter.

2. This cost is probably an underestimate. No attempt was made to determine whether uncontrolled automobiles have added costs to keep them from being more polluting than they are, nor whether some portion of the costs for natural gas combustion (activities 45a, 46a, 47a) should be allocated to pollution control. It should be emphasized, however, that the solution **x*** is independent of the base-year control method costs.

3. The solution to the Linear Programming Model (2.9) would be the same if the objective function were **Cx** minus any constant. Choosing **Cx**, to be the constant allows for a simplification in the Linear Programming Model. It is equivalent to letting the base-year costs C_{1a}, C_{2a}, C_{3a}, etc., be zero and the costs of alternative activities, such as 1d and 3b, be $(C_{1d} - C_{1a})$ and $(C_{3b} - C_{3a})$. In chapter 6, these incremental costs of abatement are called C^j.

4. The model was implemented with the MPS/360 computer program. This program is described in Beneke and Winterboer (1973).

5. It was estimated that a total of

1,650,000 tons of coal would be burned in 1975 in industrial (other than power plants), commercial, and institutional furnaces in the absence of air pollution control regulations. In the optimal solution, 380,000 tons should be low sulfur coal.

6. A union official for St. Louis gas workers claimed that there were 20 to 30 thousand gas leaks in the city. Although this charge was challenged by the company, it was a fact that 5 percent of the natural gas piped into the region was reported in gas company statements as "unaccounted for." What portion of this missing gas was attributable to meter errors, to theft by shunting meters, or to actual leakage from faulty pipes, underground storage, and necessary margins in valves and regulators could not be determined. See Kohn (1971a, pp. 616–617).

7. Although it is not shown in table 2.1, for each activity that requires conversion from coal to natural gas there is a coefficient G_j indicating the natural gas requirement per unit of abatement activity. For example, $G_{18_e} = .017$ million cubic feet, $G_{19_e} = .021$ million cubic feet, etc. The total gas requirement for 1975, 14 billion cubic feet, is the summation of each G_j times x_j^*. Model I included an availability constraint of 47 billion cubic feet of gas for air pollution control (see Kohn (1969a, pp. 579–581). However, this constraint was not binding.

8. Russell and Spofford (1977) have developed a residuals control model of the Delaware valley in which all forms of waste are accounted for and subject to control.

9. The control of air pollution without increasing liquid, solid, and thermal wastes is accomplished by a five-fold increase in the quantity of natural gas substituted for coal. This eliminates bottom ash and collectible particulate matter and obviates the need for certain electric motors.

10. The administrative costs associated with the activities x_i^* were estimated, with the help of control officials in the St. Louis region, to be approximately $500,000 a year (in 1975) for stationary source alone. This assumes that there would be no resistance to compliance.

11. This suggests that the estimated 25 percent saving in the least-cost solution as compared to the legal solution (see p. 92) may be a high estimate.

12. Gipson, Freas, and Meyer (1975) have compared linear programming and integer programming models of air quality control for Louisville, Kentucky, and have found that the total cost of abatement associated with the integer solution is approximately 4 percent higher than the divisible solution. Because the computer processing requirements for a large scale integer programming problem are excessive, Gipson et al. have experimented with algorithms that round out the linear programming solutions. These yield integer solutions with total costs averaging 9 percent more than the divisible solutions.

13. *Current Industrial Reports, Pollution Abatement: Cost and Expenditures*, MA–200 (75)–1, Department of Commerce, Bureau of the Census, Washington, D.C., 1977, p.108.

14. For example, based on projections of the utilities, it was assumed that there would be no combustion of fuel oil in power plants in 1975 (see source 13 on

table 2.1). In fact, significant quantities were burned to produce electricity.

15. It was estimated that the premium for low sulfur coal would be $1.25 per ton. In actuality, with the energy crisis this premium was closer to $6.00 per ton for large industrial firms and much more for smaller firms.

Notes to Chapter 4

1. The factors 454 and 1,000,000 convert pounds to grams and grams to micrograms. The factor 8760 converts hourly velocity to annual velocity. The values for h, s, and v are taken from data in Kohn (1972a) and Kohn and Weger (1973). The width of 100,000 meters is the approximate horizontal distance from the western boundary of St. Charles county to the eastern boundary of St. Clair county and the approximate vertical distance from the northern boundary of Madison county to the southern boundary of Jefferson county (see figure 2.1).

2. The concentration is converted from micrograms per cubic meter to parts per million according to the *ideal gas law* (see Hougen et al., 1962, pp. 51–56). One mole of sulfur dioxide weighs 64 grams and occupies .0224 cubic meters of space at 273°K. At a temperature of 287.8°K, the annual average temperature at the St. Louis CAMP station (see Environmental Data Service, 1971, p. 191), one mole of sulfur dioxide occupies (.0224) (287.8/273) (= .024) cubic meters. Proportionally, .000038 grams of sulfur dioxide would occupy .014 × 10⁻⁶ cubic meters of volume. This is equivalent to a concentration of .014 ppm. The factors used to convert $\mu g/m^3$ to ppm in this study are approximately .000843

for carbon monoxide, .001472 for total hydrocarbons, .00627 for nitrogen oxides, and .000369 for sulfur dioxide.

3. Tillman and Lee (1975) have developed a version of the box concept in which an airshed consists of contiguous boxes. This allows for a different concentration in each of the individual boxes, and therefore a range of pollutant concentrations within a single airshed. In their model, changes in the locational pattern of emissions would alter the calculated concentrations for the various component boxes.

4. Although air quality control models can be constructed for any desired interval of time, the annual model is most common. Larsen et al. (1967, p. 85) have demonstrated that "concentrations are approximately lognormally distributed for all pollutants in all cities for all averaging times." Therefore, with appropriate statistical information, an annual model can be programmed to predict short run concentrations for each of the pollutants. This procedure is an integral part of the *Air Quality Implementation Planning Program* (1970, chapter 4), which was developed by the federal government for local air quality management.

5. The stochastic analysis and related statistical data for embedding the confidence limits and probabilities within the structure of the Model are described in Kohn (1972a).

6. Parsons and Croke (1969) investigated the prospect for forecasting abatement in the Chicago area and concluded that it would be unlikely that the necessary supply of discretionary natural gas would be available when it was needed or that curtailment of in-

dustrial activities would be economically feasible. Furthermore, McFarland, Barry, and DeNardo (1969) have questioned current capabilities for predicting weather conditions conducive to air pollution buildup.

7. This use of compass sectors was adapted from work by Pooler (1961) and Clarke (1964).

8. At some distance, estimated at 22,620 meters from the CAMP staton, the pollution is uniformly distributed throughout the vertical mixing height, and equation (4.14) becomes

$$m_j = \frac{(303)\,(10^{-10})\theta_j}{v_j k_j}.$$

Note that stack height is no longer significant at this distance. The implementation of the long-term diffusion formula is described in Kohn and Weger (1973).

9. This example is documented in Kohn (1974a). The sources of climatological data are referenced in Kohn and Weger (1973).

10. See note 2. In the case of particulates, g^i converts the arithmetic average to a geometric average, since the official air quality standard for particulates is a geometric rather than an arithmetic average.

11. To keep the model simple, each composite source was disaggregated into a Missouri and an Illinois component. Each diffusion factor m_j was taken as a weighted average of the diffusion factors for the individual point or area sources in that particular category for the given state. For example, source number 78 (see table 2.1) was disaggregated into source numbers M78 and I78, which denote the quan-

tity of sulfuric acid produced, respectively, by a plant in St. Louis county and by a composite of three plants in St. Clair county. The s_j and m_j parameters for the Missouri and Illinois components are $s_{M78} = 486,100$ and $m_{M78a} = m_{M78b} = 684 \times 10^{-10}$, $s_{I78} = 623,900$ and $m_{I78a} = m_{I78b} = 2,924 \times 10^{-10}$.

12. The "corrected" values of the elements g^i are as follows: carbon monoxide, .004020; hydrocarbons .003925; nitrogen oxides, .000761; sulfur dioxide, .000375; and particulates, 1.86. The values of g^i for the gaseous pollutants indicate ppm per $\mu g/m^3$. A similar kind of correcting calibration is used in the *Air Quality Implementation Planning Program* (1970, chapter 4).

13. The diffusion factors for emissions from the three largest power plants (see table 2.1) were $m_{I38a} = m_{I38b} = m_{I38c} = m_{I38d} = 101 \times 10^{-10}$; $m_{M42a} = m_{M42b} = m_{M42c} = 112 \times 10^{-10}$; and $m_{M43a} = m_{M43b} = m_{M44a} = m_{M44b} = 74 \times 10^{-10}$. Thus a pound of pollutant from Granite City Steel Company (for whom the diffusion factor is 677×10^{-10}) had from six to nine times as much effect on the concentration at the CAMP station as did a pound of the same pollutant from one of these power plants.

14. For example, the diffusion factor for automotive emissions from Missouri was $m_{M1a} = m_{M1b} = \dots = m_{M3d} = 8,545 \times 10^{-10}$. Thus a pound of emissions from this particular area source has twelve times the impact on the CAMP station concentration as does a pound of the same pollutant emitted by Granite City Steel Company.

15. This particular study is described in greater detail in Kohn (1974a).

16. At a distance of two miles northeast,

the steel company would contribute approximately .006 ppm to the sulfur dioxide concentration at the CAMP station (see figure 4.3). This quantity is the product of the 29 million pounds of sulfur dioxide emitted by the firm (see p. 117) times the diffusion constant $5{,}136 \times 10^{-10}$ at that distance (see table 4.6) times the correction factor g^i for sulfur dioxide, which is .000375 (see note 12).

17. Relocation would also affect commuting and trucking. The costs and emissions associated with commuting are included in the pollution control model of Oron and Pines (1974) and of Schuler (1974). See also Kohn (1974b).

18. It is conceivable that high pollution levels may be economically efficient in certain industrial areas. The subject of efficient standards will be examined in the next chapter.

19. An effort was made in the case of Model I (and related models incorporating the Larsen formula) to correct for tall stacks. The emissions for source numbers 42, 43, and 44 (see table 2.1) are actually taken at 59 percent, 50 percent, and 50 percent, respectively, of their true values, to scale them to ground level equivalent emissions (see Kohn (1970a, p. 79)). Consequently, the total emission flows for 1970 (see table 4.1) corrected to ground level equivalent emissions are actually 3,108 million pounds of carbon monoxide, 939 million pounds of hydrocarbons, 385 million pounds of nitrogen oxides, 1,185 million pounds of sulfur dioxide, and 232 million pounds of particulates. This significantly reduces the predicted concentrations in column 4 for nitro-

gen oxides and sulfur dioxide, thereby improving the reliability of the Larsen formula. The projected concentrations in table 4.5 for the Larsen formula *are* based on ground-level equivalent emissions.

20. The increasing flows of emissions over time suggest that a dynamic programming model for multi-year optimization would be useful.

21. This point of view differs somewhat from that expressed by the author in testimony before the Public Works Committee of the U. S. Senate (see Kohn, 1970b). In that statement it was suggested that total emissions in an airshed should be restricted to fixed annual maxima for the various pollutants and these maxima adhered to indefinitely.

Notes to Chapter 5

1. I am grateful to Professor Deacon (1974, p. 219) for pointing out in his article that the shadow prices for Model III may be misleading. His clarification is as follows: ". . . the figure for hydrocarbons shows that the shadow price of changing the goal from 3.10 to 2.10 ppm is $15,388,000, which implies that the shadow price of going from a goal of 3.10 to 3.09 ppm is $153,880." Deacon's benefit-cost interpretation of the shadow prices includes the following: ". . for example, if the standard on particulates is 'correct,' the marginal benefit gained by reducing particulate concentrations by one unit (from 76 to 75 $\mu g/m^3$) must be approximately $239,100."

2. In Model V, the pollutant concentration for hydrocarbons is not binding and the shadow price for this concen-

tration is therefore zero. However, it is likely that concentrations less than 3.1 ppm would still cause economic damage 3. In 1965, the concentrations at the CAMP station were twice as high for particulates and four times as high for sulfur dioxide than they were twenty miles away (see Williams et al., 1967, pp. 16, 36).

4. To parametrize the interest rate, the unit capital investment requirement for each of the pollution control methods was included in the model as a coefficient. The total capital input (that is the summation of capital coefficients times corresponding activity levels) multiplied by the change in opportunity cost of capital was added to the total cost of abatement in the objective function.

5. The decrease in the interest rate caused control method 19d (for which the capital coefficient was $3.00 and which therefore decreased by $.075 in net annual cost) to replace control method 19e (see table 2.1). This had the effect of increasing the marginal cost of sulfur dioxide abatement. The optimal quantity of invested capital for pollution abatement, which includes the capital associated with the base-year level of abatement, is shown at the bottom of table 5.4. For a higher rate of interest, less capital investment is indicated. The results of Model VII are discussed at greater length in Kohn (1971c).

6. A diffusion formula such as that described in the preceding chapter would relate emissions from each source to concentrations at the various receptor locations.

7. The rate of transformation is the ratio of the marginal cost of pollution-index reduction to the marginal cost of the composite good. Teh latter is unity times a multiplier Ω that accounts for the cost of sequential rounds of derived inputs of the intermediate good. Similarly, the marginal cost of reducing the pollution index is the right-hand-side value of (5.7) times this same multiplier term. The multiplier term appears in both the numerator and denominator and cancels out; This multiplier is given by $[1 - \alpha (c_{1b} + [e_{1b}^1 w^1 + e_{1b}^2 w^2] [c_{1c} - c_{1b}]/[w_1 (e_{1b}^1 - e_{1c}^1) + [w^2(e_{1b}^2 - e_{1c}^2)])]^{-1}$. The term by which α is multiplied is the direct marginal cost of the intermediate good (that is, the marginal cost from a corresponding linear programming problem in which the sum of activity levels is constrained to αR). If each side of (5.7) is multiplied by the multiplier term, the left-hand side would represent the marginal benefits of pollution-index reduction and the right-hand side the marginal cost.

8. One of the first linear index systems was developed by Babcock (1970). In his pollution index, which he calls "Pindex," the toxicity weights are inversely proportional to State of California standards. The underlying assumption is that legal standards reflect the relative toxicity of the individual pollutants. There is a conceptual problem in using an index based on existing standards to determine optimal standards, as Babcock suggests. The theory in this book is that an optimal set of standards should be based on costs of abatement as well as benefits,

and to the extent that actual standards do reflect costs as well as benefits, they do not provide the pure toxicity weights that are required here. An index similar in concept to that of Babcock is used successfully by Dewees (1974, pp. 41–46).

9. The rate of transformation can be derived in three ways: (1) It is the shadow price of q^s in a reformulated model in which q^p is minimized. This shadow price, like the objective function, is measured in micrograms per cubic meter. (2) It is the ratio of (q^p [at c] minus q^p [at b]) to (q^s [at b] minus q^s [at c]). (3) It is the ratio of the shadow price of q^s to the shadow price of q^p from Model III. (This ratio [see table 5.2], which is ($439,100,000)/ ($239,100) $\cong$ 1840, differs somewhat from the value in table 5.7 because of rounding errors.)

10. This result would be insignificant if the production-possibility set of sulfur dioxide and particulate concentrations were extremely convex. The data in Kohn (1971d), however, suggest that there are meaningful trade-offs between the two pollutants over the range $\{q^s = .026$ ppm; $q^p = 67$ $\mu g/m^3\}$ to $\{q^s = .016$ ppm; $q^p = 100$ $\mu g/m^3\}$.

11. This particular corner solution is discussed in Page and Ferejohn (1974).

12. There are three nonzero activity levels for source 3 and two each for sources 19, 20, and 56 (see column 11 in table 2.1).

13. The Z axis extends downward, so that lower valued hyperplanes are outward and upward from the origin (which denotes preregulatory concentrations of the pollutants and the preregulatory level of Z). The pos-

sibility also exists that the hyperplane would be tangent to one of the facets, in which case there would be an infinite number of activity solutions.

14. If the model is extended to include a pollution avoidance capability, this single divisibility will vanish. This is proved in Kohn and Aucamp (1976).

15. This conclusion is valid only if the numerical input is correct. The model does appear to understate certain costs of sulfur dioxide abatement activities (i.e., flue gas desulfurization and low sulfur coal). However, the costs of the automobile control devices are also understated. On balance, the above conclusion appears to be valid.

Notes to Chapter 6

1. For the reader who is planning to construct a linear programming model for air pollution control, the idea of expressing the activity variables, x_j in units of the y_i, may occur. However, if the activity variables are expressed in units of output, there would be a cumbersome number of mutually exclusive combinations of production-abatement processes for each industrial activity. In the case of steel production, the activity set for the Linear Programming Model would include all mutually exclusive combinations of variables 5a, 5b, 11a, 11b, 11c, 46a, 51a, 53a, 53b, 68a, 68b, 69a, 69b, 70a, 70b, 71a, 71b, 85a, 85b, 92a, and 92b. The number of activity variables for steel production alone would be the product of the number of processes for each of the component activities included in the above, or 2 × 3 × 1 × 1 × 2 × 2 × 2 × 2 × 2 × 2 × 2 = 768. In the case of industries in which different firms use

different kinds of stokers (see activity variables 17a through 32c in table 2.1), the activity set would be enormous.

2. It is conceivable that there would be an interaction between **x** and the coefficients **C** and **e**. For example, the cleaner air resulting from abatement activity could lower certain costs (e.g., depreciation, maintenance, labor, etc.). Such interaction is assumed away in the Linear Programming Model.

3. The data for these examples are based on material in Kohn (1972b). Most, but not all, of the outputs y_k are measured in one dollar units at preregulatory prices.

4. The methodology for determining the feedback effect of the solution of Model I is described in Kohn (1972b).

5. An implicit assumption in this methodology is that supply curves are horizontal. Therefore, there is no loss of producers' surplus. A more realistic valuation of the loss of consumers' surplus would have been $(1/2)$ $(\mathbf{1v\hat{C}x})$.

6. The $35.3 million total cost for Model I does not include any loss of consumers' surplus. An estimate of that loss, based on the methodology used here, is $.2 million, which is approximately equal to the corresponding loss in Model IX. The reader who is interested in the consumer surplus loss from pollution abatement should see Dewees (1974, pp. 54, 75, 81, 93).

7. The empirical data for this extension of the model are described in Kohn (1975a).

8. This is the third breakdown of the $35.3 million in incremental total costs of abatement shown in this book.

9. A by-product of abatement activities 38d, 42c, 43b, and 88b is $11.9 million worth of sulfuric acid. It is assumed that this acid could be sold without affecting the existing production of sulfuric acid in the region. If, alternatively, the recovered acid were to crowd out an equal-valued quantity of commercial sulfuric acid production, the feedback impact would be dampened and the total cost of abatement would increase by $300,000 rather than $800,000 (see Kohn, 1975a, pp. 245, 258, 259).

10. In the original research, a different classification of economic activities was used for each of the feedback studies. Accordingly, two different **a** matrices were required. Researchers who are planning similar models are advised to select a single classification of economic activities, preferably one for which input-output multipliers are available.

11. Because the individual abatement activities are constrained by source magnitudes, air pollution control cannot be treated as a constant cost industry as Leontief (1970) has assumed.

Notes to Chapter 7

1. The set of abatement activities that polluters would voluntarily implement if they were charged emission fees π is not necessarily identical to the corresponding least-cost set of activity levels $\mathbf{x}^*$. Some of the economic costs of abatement are not private accounting costs and therefore would not influence decision making. In addition, there are likely to be divisibilities in the activity set, as with source numbers 3, 19, 20, and 56 in the Linear Programming Model (see column 11 of table 2.1). Maler (1974, p. 208) has shown that

pollution fees will not necessarily induce the correct combination of abatement activities when there are divisibilities in the optimal solution. However, Kohn and Aucamp (1976) have demonstrated that when households can take measures to reduce their exposure to pollution, the optimal level of pollution (or of a combined pollution index) implies an integral abatement solution void of divisibilities.

2. The interrelationship of Least-Cost Standards and Pigouvian Fees is discussed by Baumol and Oates (1971, p. 44) and by Abrams and Barr (1974).

3. The political pressures that prevent regulatory agencies from adopting least-cost standards are described in Ruff (1970).

4. This is illustrated by the case of an enamelling firm in Illinois, which could meet standards only by installing a baghouse collector. Expecting that this would be very costly, the firm delayed compliance and paid some substantial fines. When it finally did install the collection equipment, it was discovered that the volume of recovered aerosols was greater, and its quality superior, to what had been expected, and that the material could be reused profitably in undercoating applications.

5. For contrary views, see Solow (1972), Baumol (1972, p. 319), and Weidenbaum (1975, p. 110).

6. The entries in column (6) are based on allowable flows carried to one decimal point rather than rounded out. See Kohn (1977a).

7. The same conclusion prevails if both inequalities in (7.5) are reversed. This also applies to the following discussion of (7.6) and (7.7). This methodology for evaluating the effect of Least-Cost Standards on relative price distortion was developed in Kohn (1977a). However, the technique used here has been further refined.

8. This research is described in Kohn (1974a). It is interesting that, except for a certain lumpiness, the total emission fees paid by the steel company should exactly equal the abatement costs which its location imposes on all other pollution sources in the airshed. In addition, the pollutant fees should also equal the dollar equivalent of the marginal disutility of the respective pollutants.

9. It is assumed in Model V that the output of the steel company is the same at each location. In fact, as the Pigouvian fees and abatement outlays increase, there would be price rises; this would reduce the quantities demanded and therefore slow the rate of increase in column 3 of table 7.5.

10. This point was first made by Dorfman and Dorfman (1972, p. xxxiii).

11. Freeman (1972) has shown that this is the case in Kansas City, St. Louis, and Washington, D.C.

12. Yan et al. (1975) based their research on the Philadelphia region. With their model, they were also able to determine the extent to which pollution control costs are exported to other regions.

13. This assumes that all of the allowable emissions (see table 6.2) would be taxed at the rates in (7.16), and ignores administrative and enforcement expenditures.

14. Vickery (1967) has optimistically noted that "To the extent that effluent fees provide a flow of revenue into the public treasury, this makes possible the

abatement of other taxes and impositions that have a baneful effect on the efficiency with which the economy operates." The fact that Pigouvian Fees would eliminate inefficient forms of taxation reinforces the high rankings for efficiency in table 7.1.

15. Emerson and Leitner (1976) have shown that employment in certain industries may not increase as rapidly in regions characterized by relatively stringent pollution regulations.

16. This is the increase in employment net of the direct job loss, $\mathbf{wx}$, in (7.18).

17. This research is described in Kohn (1973).

18. It would be singnificant if the control method solution, $\mathbf{x}^D$, to the above model in which D is minimized could be related to the legal regulations in the St. Louis airshed (that is $\mathbf{x}^r$). This might, in part, explain why conventional abatement regulations deviate from the least-cost solution $\mathbf{x}^*$ of Model I. In one respect at least there is a similarity between $\mathbf{x}^D$ and $\mathbf{x}^r$; in both there is greater emphasis on low sulfur coal, and less emphasis on natural gas, than in the least-cost solution. It is conceivable that the regulatory agencies were concerned with a potential loss of jobs in the coal industry if this fuel were banned for particular types of stokers. Except for this one similarity, there was very little in solution $\mathbf{x}^D$ that would explain any of the differences between the regulatory solution $\mathbf{x}^r$ and the least-cost solution $\mathbf{x}^*$.

19. This may be related to the employment effect. If too much of the cost burden falls on a particular industry, there is the fear that firms in that

industry would move away from the region.

Notes to Chapter 8

1. The emphasis in Washington has been on a model incorporating a diffusion formula (see *Air Quality Implementation Planning Program*, 1970). Although this computer model is available for use by local agencies it is very complex and is not an economic optimization model.

2. The usefulness of the Linear Programming Model and its adaptability were recently demonstrated by students in the author's economics seminar at Southern Illinois University. Each used Model I to test some aspect of pollution control. One student increased the cost of flue gas desulfurization to $10.00 per top of coal (based on a current estimate of 4 mills per kilowatt hour) and found that, although the total cost of abatement almost doubled, this process was still efficient for the major power plants. Another student, concerned with the scarcity of natural gas, disallowed conversions of industrial, commercial, and utility furnaces from coal to natural gas. This saved 6.4 billion cubic of gas and increased the total cost of abatement by $1.7 million. This represented a cost of $.27 per thousand cubic feet, which appeared quite modest. One student, interested in the President's proposal to give a rebate to purchasers of automobiles using less gasoline per mile, found that a 50 percent fuel saving, in terms of the magnitude of source number 3, would reduce the total cost of abatement in Model I by $21.00 per car per year. Another student, experienced in the grain industry, discovered a weakness in

the data. By disaggregating abatement activity 72b, he showed that certain grain cleaning devices became optimal. Another student found that if the benzo-(a)pyrene standard were reduced from .001 to .00075 $\mu g/m^3$, the shadow price for that pollutant, converted from milligrams to pounds, would be $16,000 per pound. This information proved to be valuable and was used in Kohn (1977b).

3. See *Environmental Quality—1976*, pp. 247–253.

4. This is proved by Berglas (1977) with a theoretical model.

APPENDIX:
COMPUTER PRINTOUT FOR
MODEL I USING DATA
IN TABLE 2.1

SECTION 1 - ROWS

NUMBER	...ROW..	AT	...ACTIVITY...	SLACK ACTIVITY	..LOWER LIMIT.	..UPPER LIMIT.	.DUAL ACTIVITY
1	COST	PS	46525242.2638	46525242.2638-	NONE	NONE	1.00000
2	S1	EQ	330700.00000	.	330700.00000	330700.00000	65.63964-
3	S2	EQ	180800.00000	.	180800.00000	180800.00000	65.63964-
4	S3	EQ	625500.00000	.	625500.00000	625500.00000	46.36387-
5	S4	EQ	7252.00000	.	7252.00000	7252.00000	86.57237-
6	S5	EQ	69350.00000	.	69350.00000	69350.00000	86.57237-
7	S6	EQ	136500.00000	.	136500.00000	136500.00000	10.57825-
8	S7	EQ	14700.00000	.	14700.00000	14700.00000	.47373-
9	S8	EQ	2100.00000	.	2100.00000	2100.00000	1.89342-
10	S9	EQ	52500.00000	.	52500.00000	52500.00000	3.20575-
11	S10	EQ	4200.00000	.	4200.00000	4200.00000	6.01632-
12	S11	EQ	39700.00000	.	39700.00000	39700.00000	40.63935-
13	S12	EQ	158800.00000	.	158800.00000	158800.00000	39.74745-
14	S13	PS	.	.	.	.	.
15	S14	EQ	11540.00000	.	11540.00000	11540.00000	25.43864-
16	S15	EQ	150100.00000	.	150100.00000	150100.00000	25.43864-
17	S16	EQ	8017.00000	.	8017.00000	8017.00000	25.43864-
18	S17	EQ	580.00000	.	580.00000	580.00000	7.55434-
19	S18	EQ	2400.00000	.	2400.00000	2400.00000	8.09434-
20	S19	FO	173000.00000	.	173000.00000	173000.00000	10.07246-
21	S20	EQ	358800.00000	.	358800.00000	358800.00000	10.17246-
22	S21	EQ	97600.00000	.	97600.00000	97600.00000	10.15246-
23	S22	EQ	114300.00000	.	114300.00000	114300.00000	11.58670-
24	S23	EQ	444000.00000	.	444000.00000	444000.00000	10.79673-
25	S24	EQ	110000.00000	.	110000.00000	110000.00000	9.66943-
26	S25	EQ	29000.00000	.	29000.00000	29000.00000	3.45398-
27	S26	EQ	35000.00000	.	35000.00000	35000.00000	8.86733-
28	S27	EQ	63500.00000	.	63500.00000	63500.00000	9.71733-
29	S28	FO	8000.00000	.	8000.00000	8000.00000	9.99992-
30	S29	EQ	17000.00000	.	17000.00000	17000.00000	8.28992-
31	S30	EQ	28000.00000	.	28000.00000	28000.00000	8.36997-
32	S31	EQ	15000.00000	.	15000.00000	15000.00000	9.30942-
33	S32	EQ	1120.00000	.	1120.00000	1120.00000	3.15724-
34	S33	EQ	428000.00000	.	428000.00000	428000.00000	3.91231-
35	S34	EQ	32000.00000	.	32000.00000	32000.00000	3.97720-
36	S35	EQ	24000.00000	.	24000.00000	24000.00000	8.05941-
37	S36	EQ	60000.00000	.	60000.00000	60000.00000	7.69358-
38	S37	EQ	44000.00000	.	44000.00000	44000.00000	7.01158-
39	S38	EQ	1000000.00000	.	1000000.00000	1000000.00000	6.91769-
40	S39	EQ	200000.00000	.	200000.00000	200000.00000	11.11619-
41	S40	EQ	132000.00000	.	132000.00000	132000.00000	10.39895-
42	S41	EQ	730000.00000	.	730000.00000	730000.00000	7.24767-
43	S42	EQ	3000160.00000	.	3000160.00000	3000160.00000	4.49295-
44	S43	EQ	3430000.00000	.	3430000.00000	3430000.00000	3.87484-
45	S44	EQ	2070000.00000	.	2070000.00000	2070000.00000	1.55030-
46	S45	FO	10500.00000	.	10500.00000	10500.00000	139.25826-
47	S46	EQ	123000.00000	.	123000.00000	123000.00000	82.34756-
48	S47	EQ	112500.00000	.	112500.00000	112500.00000	50.13871-
49	S48	EQ	38000.00000	.	38000.00000	38000.00000	53.13470-
50	S49	EQ	7262.00000	.	7262.00000	7262.00000	98.39667-
51	S50	EQ	26176.00000	.	26176.00000	26176.00000	147.18000-
52	S51	EQ	17974.00000	.	17974.00000	17974.00000	59.47769-
53	S52	EQ	357000.00000	.	357000.00000	357000.00000	2.14995-
54	S53	EQ	50000.00000	.	50000.00000	50000.00000	1.01051-
55	S54	FO	33600.00000	.	33600.00000	33600.00000	2.43688-
56	S55	EQ	16500.00000	.	16500.00000	16500.00000	2.40332-
57	S56	FO	519000.00000	.	519000.00000	519000.00000	11.16000-
58	S57	EQ	455000.00000	.	455000.00000	455000.00000	.93330-
59	S58	FO	706000.00000	.	706000.00000	706000.00000	.55800-
60	S59	EQ	30600.00000	.	30600.00000	30600.00000	4.10000-
61	S60	EQ	3400.00000	.	3400.00000	3400.00000	40.00000-
62	S61	FO	97000.00000	.	97000.00000	97000.00000	1.04662-
63	S62	EQ	36000.00000	.	36000.00000	36000.00000	.26145-
64	S63	FO	20300.00000	.	20300.00000	20300.00000	.82130-
65	S64	EQ	142100.00000	.	142100.00000	142100.00000	.72058-
66	S65	EQ	40600.00000	.	40600.00000	40600.00000	1.46162-
67	S66	EQ	94300.00000	.	94300.00000	94300.00000	1.42230-
68	S67	EQ	25600.00000	.	25600.00000	25600.00000	.40240-
69	S68	EQ	1000000.00000	.	1000000.00000	1000000.00000	.32243-
70	S69	EQ	2750000.00000	.	2250000.00000	2250000.00000	.17390-
71	S70	EQ	850000.00000	.	850000.00000	850000.00000	.26550-
72	S71	EQ	1417500.00000	.	1417500.00000	1417500.00000	.16715-
73	S72	EQ	2400000.00000	.	2400000.00000	2400000.00000	42.89715-
74	S73	EQ	54604.00000	.	137606.00000	137606.00000	3.95826-
75	S74	EQ	137606.00000	.	200000.00000	200000.00000	.58324-
76	S75	EQ	200000.00000	.	100000.00000	100000.00000	1.44755-
77	S76	EQ	100000.00000	.	125000.00000	125000.00000	.14399-
78	S77	EQ	125000.00000	.	125000.00000	125000.00000	

| --- | --- | --- | --- | --- | --- | --- | --- |
79	S78	EO	1110000.00000	.	1110000.00000	1110000.00000	1.10968-
80	S79	FO	410000.00000	.	410000.00000	410000.00000	.51968-
81	S80	EO	3760000.00000	.	3760000.00000	3760000.00000	.17810-
82	S81	FO	2455000.00000	.	2455000.00000	2455000.00000	.22196-
83	S82	FO	9285000.00000	.	9285000.00000	9285000.00000	.16092-
84	S83	EO	6250.00000	.	6250.00000	6250.00000	1.35593-
85	S84	EO	461000.00000	.	461000.00000	461000.00000	.05487-
86	S85	EQ	1326000.00000	.	1326000.00000	1326000.00000	.15496-
87	S86	EO	440000.00000	.	440000.00000	440000.00000	1.45169-
88	S87	EQ	1450.00000	.	1450.00000	1450.00000	93.71000
89	S88	EO	218000.00000	.	218000.00000	218000.00000	2.36877-
90	S89	EQ	432000.00000	.	432000.00000	432000.00000	3.01394-
91	S90	EG	4000.00000	.	4000.00000	4000.00000	219.07415-
92	S91	EO	2600000.00000	.	2600000.00000	2600000.00000	.34394-
93	S92	FO	24800.00000	.	24800.00000	24800.00000	49.51691-
94	S93	EO	5000.00000	.	5000.00000	5000.00000	49.51691-
95	S94	EO	4820.00000	.	4820.00000	4820.00000	49.51691-
96	S95	EO	868000.00000	.	868000.00000	868000.00000	2.63565-
97	S96	EO	226000.00000	.	226000.00000	226000.00000	2.72565-
58	S97	FO	64000.00000	.	64000.00000	64000.00000	2.63565-
99	CO	UL	2335162803.00	.	NONE	2335162803.00	.00042h
100	MC	UL	994455118.800	.	NONE	994455118.000	.02476
101	NO	UL	303532644.000	.	NONE	303532644.000	.32639
102	SO	UL	430424669.000	.	NONE	430424669.000	.02193
103	PM	UL	135802509.000	.	NONE	135802509.000	.07748
104	MP	RS	5720685648.500	60987281.5002	NONE	633055870.000	.

| --- | --- | --- | --- | --- | --- | --- | --- |
105	X1A	RS	333700.00000	.	.	NONE	.
106	X1B	LL	.	26.30000	.	NONE	14.57317
107	X1C	LL	.	53.00000	.	NONE	20.49423
108	X1D	LL	.	76.30000	.	NONE	35.06741
109	X2A	RS	180800.00000	.	.	NONE	.
110	X2B	LL	.	19.50000	.	NONE	7.77317
111	X2C	LL	.	50.00000	.	NONE	20.49423
112	X2D	LL	.	69.50000	.	NONE	28.26741
113	X3A	LL	.	.	.	NONE	19.25577
114	X3B	PS	79862.11905	10.25000	.	NONE	.
115	X3C	BS	519753.26538	28.71000	.	NONE	.
116	X3C	BS	25884.61467	30.76000	.	NONE	.
117	X4A	RS	7252.00000	.	.	NONE	.
118	X4B	LL	.	14.36000	.	NONE	.
119	X5A	RS	69350.00000	.	.	NONE	99.12177
120	X5B	LL	.	28.72000	.	NONE	.
121	X6A	BS	136530.00000	.	.	NONE	26.90949
122	X7A	BS	14700.00000	.	.	NONE	.
123	X8A	RS	2100.00000	.	.	NONE	.
124	X9A	RS	52507.00000	.	.	NONE	.
125	X10A	RS	4200.00000	.	.	NONE	.
126	X11A	LL	.	.	.	NONE	.
127	X11B	BS	39700.00000	2.50000	.	NONE	.09675
128	X11C	LL	.	33.50000	.	NONE	30.65572
129	X12A	BS	158800.00000	.	.	NONE	.
130	X12B	LL	.	5.00000	.	NONE	3.39232
131	X13A	LL	.	.	.	NONE	40.72631
132	X14A	BS	11543.00000	.	.	NONE	.
133	X15A	BS	150100.00000	.	.	NONE	.
134	X16A	RS	8017.00000	.	.	NONE	.
135	X17A	LL	.	.	.	NONE	5.46800
136	X17B	LL	.	1.62000	.	NONE	3.79503
137	X17C	LL	.	2.41000	.	NONE	6.62800
138	X17D	LL	.	4.03000	.	NONE	4.95503
139	X17F	RS	580.00000	6.16000	.	NONE	.
140	X18A	LL	.	.63000	.	NONE	4.13522
141	X18B	LL	.	1.85000	.	NONE	3.48503
142	X18C	LL	.	3.04000	.	NONE	5.26522
143	X18E	LL	.	4.26000	.	NONE	4.64503
144	X18F	RS	2400.00000	6.70000	.	NONE	.
145	X19A	LL	.	.	.	NONE	2.95037
146	X19B	LL	.	.73000	.	NONE	.
147	X19C	LL	.	1.25000	.	NONE	2.95037
148	X19D	RS	66970.08444	1.98000	.	NONE	.
149	X19E	RS	106029.91556	8.34000	.	NONE	.
150	X20A	LL	.	.28000	.	NONE	1.67759
151	X20B	BS	44400.50038	.83000	.	NONE	.
152	X20C	LL	.	1.53000	.	NONE	1.67759
153	X20D	RS	314399.49962	2.08000	.	NONE	.
154	X20E	LL	.	8.54000	.	NONE	.13000
155	X21A	LL	.	.45000	.	NONE	.22111
156	X21B	BS	57600.00000	.81000	.	NONE	.
157	X21C	LL	.	1.70000	.	NONE	.22111
158	X21D	LL	.	2.06000	.	NONE	.
159	X21E	LL	.	8.62000	.	NONE	.23000
160	X22A	LL	.	.43000	.	NONE	2.71117
161	X22B	BS	114000.00000	1.19000	.	NONE	.
162	X22C	LL	.	1.68000	.	NONE	2.71117
163	X22D	LL	.	2.44000	.	NONE	1.16532
164	X22F	LL	.	10.53000	.	NONE	1.85574
165	X23A	LL	.	.90000	.	NONE	.49176

NUMBER	.COLUMN.	AT	...ACTIVITY...	..INPUT COST..	..LOWER LIMIT.	..UPPER LIMIT.	.REDUCED COST.
166	X23B	BS	444300.00000	1.40000	.	NONE	.
167	X23C	LL	.	2.15000	.	NONE	1.60750
168	X23D	LL	.	2.65000	.	NONE	1.16532
169	X23E	LL	.	10.82000	.	NONE	1.93574
170	X24A	LL	.	.90000	.	NONE	1.59437
171	X24B	BS	110000.00000	4.17000	.	NONE	.
172	X24C	LL	.	10.82000	.	NONE	3.03835
173	X25A	LL	.	.90000	.	NONE	7.83458
174	X25B	BS	29300.00000	.48000	.	NONE	.
175	X26A	LL	.	.45000	.	NONE	4.60550
176	X26B	LL	.	.77000	.	NONE	2.21365
177	X26C	LL	.	1.70000	.	NONE	4.60550
178	X26D	LL	.	2.02000	.	NONE	2.21365
179	X26E	BS	35300.00000	7.18000	.	NONE	.
180	X27A	LL	.	.56000	.	NONE	2.54105
181	X27B	LL	.	.77000	.	NONE	2.36365
182	X27C	LL	.	1.81000	.	NONE	2.54135
183	X27D	LL	.	2.02000	.	NONE	2.36365
184	X27E	BS	63500.00000	7.33300	.	NONE	.
185	X28A	LL	.	.	.	NONE	11.12144
186	X28B	LL	.	.64000	.	NONE	2.44429
187	X28C	LL	.	1.25000	.	NONE	11.12114
188	X28D	LL	.	1.89000	.	NONE	2.44429
189	X28E	BS	8300.00000	6.48000	.	NONE	.
190	X29A	LL	.	.28000	.	NONE	8.18966
191	X29B	LL	.	.76000	.	NONE	2.37429
192	X29C	LL	.	1.53000	.	NONE	8.18966
193	X29D	LL	.	2.01000	.	NONE	2.37429
194	X29E	BS	17000.00000	6.67000	.	NONE	.
195	X30A	LL	.	.45000	.	NONE	3.74699
196	X30B	LL	.	.77000	.	NONE	2.30429
197	X30C	LL	.	1.70000	.	NONE	3.74699
198	X30D	LL	.	2.32000	.	NONE	2.30429
199	X30E	BS	28700.00000	6.75000	.	NONE	.
200	X31A	LL	.	.56000	.	NONE	2.40610
201	X31B	LL	.	.77000	.	NONE	2.36429
202	X31C	LL	.	1.81000	.	NONE	2.40610
203	X31D	LL	.	2.02000	.	NONE	2.36429
204	X31F	BS	15300.00000	6.69000	.	NONE	.
205	X32A	LL	.	.	.	NONE	7.54114
206	X32B	LL	.	2.41000	.	NONE	8.68830
207	X32C	BS	1120.00000	2.47000	.	NONE	.
208	X33A	LL	.	.	.	NONE	5.61784
209	X33B	BS	428000.00000	3.06000	.	NONE	.
210	X34A	LL	.	.	.	NONE	3.23351
211	X34B	BS	32000.00000	2.47000	.	NONE	.
212	X35A	BS	24300.00000	.63000	.	NONE	.
213	X35B	LL	.	1.85000	.	NONE	.25148
214	X35C	LL	.	3.06000	.	NONE	1.16333
215	X35D	LL	.	4.26000	.	NONE	1.41148
216	X35E	LL	.	11.76000	.	NONE	4.75908
217	X36A	LL	.	.	.	NONE	1.67296
218	X36B	BS	60300.00000	1.62000	.	NONE	.
219	X36C	LL	.	2.41000	.	NONE	2.83294
220	X36D	LL	.	4.33300	.	NONE	1.16030
221	X36E	LL	.	11.21000	.	NONE	4.57498
222	X37A	LL	.	.	.	NONE	2.35486
223	X37B	LL	.	1.62000	.	NONE	.63190
224	X37C	LL	.	2.41000	.	NONE	3.51456
225	X37D	LL	.	4.33000	.	NONE	1.84193
226	X37E	BS	44000.00000	6.16000	.	NONE	.
227	X38A	LL	.	.16000	.	NONE	4.22017
228	X38B	LL	.	.79000	.	NONE	3.23851
229	X38C	LL	.	1.21000	.	NONE	4.02012
230	X38D	BS	1000000.00000	1.43000	.	NONE	.
231	X38E	LL	.	9.36000	.	NONE	5.69233
232	X39A	LL	.	.30000	.	NONE	.49176
233	X39B	BS	200000.00000	.80000	.	NONE	.
234	X39C	LL	.	1.35000	.	NONE	1.40753
235	X39D	LL	.	1.85000	.	NONE	.96532
236	X40A	BS	132300.00000	.35000	.	NONE	.
237	X40B	LL	.	1.40000	.	NONE	.90334
238	X40C	LL	.	6.40000	.	NONE	5.36816
239	X40D	LL	.	7.45000	.	NONE	6.33426
240	X40F	LL	.	9.37000	.	NONE	2.20096
241	X41A	LL	.	.50000	.	NONE	2.61652
242	X41B	LL	.	1.55000	.	NONE	3.57874
243	X41C	BS	730000.00000	1.70000	.	NONE	.
244	X42A	LL	.	.50000	.	NONE	.75763
245	X42B	LL	.	1.55000	.	NONE	1.30292
246	X42C	BS	3000160.00000	1.77000	.	NONE	.
247	X43A	LL	.	.55000	.	NONE	1.26629
248	X43B	BS	3430000.00000	1.13000	.	NONE	.
249	X44A	LL	.	.55000	.	NONE	3.59012
250	X44B	BS	2070000.00000	1.55000	.	NONE	.
251	X45A	BS	10500.00000	.	.	NONE	.
252	X46A	BS	123000.00000	.	.	NONE	.
253	X47A	BS	112500.00000	.	.	NONE	.
254	X48A	BS	38000.00000	.	.	NONE	.
255	X49A	BS	7267.00000	.	.	NONE	.
256	X49B	LL	.	67.03000	.	NONE	50.33333
257	X50A	BS	26176.00000	67.00000	.	NONE	.
258	X51A	BS	17974.00000	.	.	NONE	.
259	X52A	BS	357000.00000	.	.	NONE	.
260	X52B	LL	.	.87000	.	NONE	.34313

NUMBER	COLUMN	AT	...ACTIVITY...	..INPUT COST..	..LOWER LIMIT.	..UPPER LIMIT.	.REDUCED COST.
261	X53A	AS	50000.00000	.	.	NONE	.
262	X53B	LL	.	1.00000	.	NJNE	.73656
263	X54A	AS	33600.00000	.	.	NUNE	.
264	X54B	LL	.	2.82000	.	NONE	.38312
265	X55A	BS	16500.00000	.	.	NONE	.
266	X55B	LL	.	2.82000	.	XUNF	.41698
267	X56A	AS	400408.68811	.	.	NUNE	.
268	X56B	AS	118591.31189	11.16000	.	NUNE	.
269	X57A	LL	.	.	.	NONE	11.23000
270	X57B	AS	455000.00000	.93000	.	NONE	.
271	X58A	AS	706000.00000	.	.	NONE	.
272	X58B	LL	.	.72000	.	NONE	.16200
273	X59A	LL	.	.	.	NONE	7.07477
274	X59B	AS	30600.00000	4.10000	.	NONE	.
275	X60A	LL	.	.	.	NONE	51.17477
276	X60B	BS	3400.00000	40.00000-	.	NONE	.
277	X61A	BS	97000.00000	.23000	.	NONE	.
278	X62A	AS	36300.00000	.06000	.	NONE	.
279	X62B	LL	.	.39000	.	NONE	.12630
280	X63A	BS	20300.00000	.	.	NONE	.
281	X63B	LL	.	2.90000	.	NONE	2.09419
282	X64A	BS	142100.00000	.	.	NONE	.
283	X64B	LL	.	1.55000	.	NONE	.84141
284	X65A	BS	40600.00000	1.45000	.	NONE	.
285	X66A	BS	94000.00000	.	.	NONE	.
286	X66B	LL	.	1.10000	.	NONE	.13923
287	X66C	LL	.	3.40000	.	NONE	2.08354
288	X66D	LL	.	7.20000	.	NONE	5.94483
289	X67A	AS	25600.00300	.	.	NONE	.
290	X68A	LL	.	.	.	NONE	.49891
291	X68B	BS	1000000.00000	.31000	.	NONE	.
292	X69A	BS	2250000.00000	.14000	.	NONE	.
293	X69B	LL	.	.28000	.	NONE	.12140
294	X70A	LL	.	.02000	.	NONE	.02564
295	X70B	BS	850000.00000	.25000	.	NONE	.
296	X71A	LL	.	.07000	.	NONE	.32045
297	X71B	AS	1417500.00000	.16000	.	NONE	.
298	X72A	BS	2400000.00000	.04000	.	NONE	.
299	X72B	LL	.	.49000	.	NONE	.01998
300	X73A	LL	.	.	.	NONE	58.58650
301	X73B	AS	54604.00000	2.30000	.	NONE	.
302	X73C	LL	.	17.00000	.	NONE	70.98417
303	X73D	LL	.	15.30000	.	NONE	12.39760
304	X73E	LL	.	206.00000	.	NONE	254.84973
305	X73F	LL	.	238.30000	.	NONE	196.26316
306	X73G	LL	.	220.00000	.	NONE	267.24713
307	X73H	LL	.	225.30000	.	NONE	208.66076
308	X74A	AS	137606.00000	6.71000	.	NONE	.
309	X75A	LL	.	.	.	NONE	.57898
310	X75B	AS	200000.00000	.56000	.	NONE	.
311	X76A	LL	.	.	.	NONE	16.50394
312	X76B	BS	100000.00000	.55000	.	NONE	.
313	X77A	LL	.	.04000	.	NONE	.51886
314	X77B	AS	125000.00000	.11000	.	NONE	.
315	X78A	LL	.	.16000	.	NONE	.06816
316	X78B	BS	1110000.00000	.98000	.	NONE	.
317	X79A	LL	.	.16000	.	NONE	.65816
318	X79B	BS	410000.00000	.39000	.	NONE	.
319	X80A	BS	3760000.00000	.10000	.	NONE	.
320	X80B	LL	.	.21000	.	NONE	.04754
321	X80C	LL	.	.27000	.	NONE	.09492
322	X81A	BS	2455000.00000	.07000	.	NONE	.
323	X81B	LL	.	.20000	.	NONE	.01300
324	X81C	LL	.	.23000	.	NONE	.31976
325	X82A	BS	9285000.00000	.14000	.	NONE	.
326	X82B	LL	.	.17000	.	NONE	.01605
327	X83A	BS	6250.00000	.	.	NONE	.
328	X84A	AS	961300.00000	.02000	.	NONE	.
329	X85A	BS	1326000.00000	.	.	NONE	.
330	X85B	LL	.	.23000	.	NONE	.12153
331	X86A	LL	.	.	.	NONE	.56293
332	X86B	LL	.	.01000	.	NONE	.52634
333	X86C	LL	.	.04000	.	NONE	.06346
334	X86D	LL	.	.05000	.	NONE	.02597
335	X86E	BS	440000.00000	.15000	.	NONE	.
336	X86F	LL	.	.39000	.	NONE	.13153
337	X87A	LL	.	.	.	NONE	353.51399
338	X87B	BS	1450.00000	90.70000-	.	NONE	.
339	X88A	LL	.	.	.	NONE	15.33005
340	X88B	BS	218000.00000	1.14000-	.	NONE	.
341	X89A	BS	432000.00000	.86000	.	NONE	.
342	X89B	LL	.	4.20000	.	NONE	1.40141
343	X90A	LL	.	3.00000	.	NONE	171.33340
344	X90B	AS	4000.00000	6.00000	.	NONE	.
345	X91A	BS	2600000.00000	.	.	NONE	.
346	X92A	AS	24800.00000	.	.	NONE	.
347	X92B	LL	.	108.00000	.	NONE	63.43478
348	X93A	BS	5000.00000	.	.	NONE	.
349	X94A	BS	4820.00000	.	.	NONE	.
350	X95A	AS	668000.00000	.14000	.	NONE	.
351	X96A	LL	.	.	.	NONE	8.80684
352	X96B	BS	226000.00000	.23000	.	NONE	.
353	X97A	LL	.	.	.	NONE	8.89684
354	X97B	BS	64000.00000	.14000	.	NONE	.

GLOSSARY OF
MATHEMATICAL SYMBOLS

a_{ik} Quantity of pollution source i per unit of output of good, or economic sector, k. This coefficient is an element of the $m \times t$ matrix, $\mathbf{a}$.

b^i The background ambient air concentration of the ith pollutant. The $p \times 1$ vector of background concentrations is $\mathbf{b}$.

$\mathbf{B}$ The basis matrix in the solution of the Linear Programming Model.

c_j Unit cost of production associated with the jth production-abatement process. These are elements of the $1 \times t$ vector $\mathbf{c}$. Alternatively, c_k is used to represent the unit cost, given the base-year level of abatement, of good k.

C_j Cost of pollution abatement only, associated with one unit of activity j. These are elements of a $1 \times n$ vector $\mathbf{C}$.

C^j The incremental cost of abatement, over and above the base-year cost C_j for activity j. The $n \times n$ diagonal matrix containing this element and the zero-valued C^j is $\hat{\mathbf{C}}$. Only private or accounting costs are included in the C^j.

d^i Diffusion constant for the ith pollutant in the Stochastic formula

D Total job displacement as a consequence of incremental abatement activity.

e_j^i Pounds of pollutant i emitted per unit of activity j. In certain models the emission rate is measured in units of the resulting contribution to an ambient air concentration. In the One-Pollutant model, the emission rate is simply e_j.

The $p \times n$ matrix of emission factors e_j^i is $\mathbf{e}$. When the diffusion formula is used the dimensions of this matrix expand to $\gamma p \times N$, where γ is the number of receptor stations.

E_j^i The legal emission rate of pollutant i per unit of pollution source j. This coefficient is an element of the $p \times m$ matrix $\mathbf{E}$.

f^i The total flow of emissions of pollutant i from all sources. The $p \times 1$ vector of emission flows is $\mathbf{f}$. The allowable flow $\hat{f}^i$ is an element of the vector $\hat{\mathbf{f}}$. The projected flows in the absence of pollution control standards is $\mathbf{f}_*$ and the set of Pareto optimal flows is $\mathbf{f}^*$.

G_j Natural gas requirement per unit of abatement activity j.

g^i A factor that converts grams per cubic meter into parts per million for the gaseous pollutants and converts the arithmetic mean for particulates to a geometric mean. The $p \times p$ diagonal matrix whose nonzero element is g^i is $\mathbf{g}$.

h_{kj} Quantity of good k (or of the output of economic sector k) used as an input per unit of abatement activity j. These are interpreted as incremental inputs over and above the base-year requirements. The $t \times n$ matrix of elements h_{kj} is $\mathbf{h}$.

H_j Effective stack height of pollution source for which the jth abatement activity is defined.

k_j Distance in meters from the receptor station of the source for which the kth abatement activity is defined.

L_{ij} A Leontief multiplier denoting the increase in the value of output of sector i resulting from a dollar increase in sales of sector j. This multiplier is an element of the $t \times t$ matrix **L**.

m^i A constant of proportionality relating the ambient air concentration (net of the background level) of pollutant i to the total annual emission flow of that pollutant. The $p \times p$ diagonal matrix whose nonzero element is m^i is **m**.

m_j The contribution in micrograms per cubic meter of the jth pollution source per pound of emissions. This coefficient, which is the same for each pollutant, is the nonzero element of an $N \times N$ diagonal matrix **M**.

p_i Competitive market selling price of good i in the absence of air pollution control. The vector of market prices is **p**.

p_i' Selling price of good i, equal to the marginal cost of production-abatement under regulatory standards.

p_i^* Selling price of good i, equal to the marginal social costs of production. The vector of such prices is **p***.

P_{ik} Proportion of source level i used in industry k.

q^i Ambient air concentration of pollutant i. In the one-pollutant model there is no superscript. The definitive superscripts are c (carbon monoxide), h (hydrocarbons), n (nitrogen oxides), s (sulfur dioxide), and p (particulates). The $p \times 1$ vector of ambient air concentrations is **q**. The legal standard is $\hat{q}^i$, which is an element of the vector $\hat{\mathbf{q}}$. The concentrations in the absence of

pollution control are $\mathbf{q}_s$, and the set of Pareto optimal concentrations is **q***.

Q An overall measure of pollution severity, such as a pollution index number.

R Quantity of labor input in the region, usually interpreted as a fixed quantity.

s_i Quantity of intermediate good i or magnitude of pollution source i. This quantity is the typical element of the $m \times 1$ vector **s**.

t_{ij} The value of the change in distortion of the ijth price ratio.

T_{ij} The distortion of the ijth price ratio.

u_{ij} Coefficient of the jth activity variable, which is unity when that variable is defined for the ith pollution source and zero otherwise. This coefficient is the typical element of the $m \times n$ distributive matrix **u**.

$\bar{U}^i$ A specific level of utility of the ith household. The utility function is $U^i(\cdot)$. The marginal rate of substitution in consumption between goods j and k is U_j^i/U_k^i, and between good-one and the level of air pollution U_1^i/U_4^i.

v Annual average wind velocity through the airshed. The annual average velocity of winds from the source for which the jth abatement activity is defined, toward the receptor station, is v_j.

v_{kj} The change in output of good k (or sales of sector k) per unit of the jth control method activity as a consequence of the feedback of abatement cost on quantity demanded. This is the typical element of the $t \times n$ matrix **v**.

w^i The relative toxicity weight of the ith pollutant.

w_j The direct loss in employment associated with one unit of abatement activity j. This coefficient is an element of the $1 \times n$ vector $\mathbf{w}$.

W^k The number of employees in the St. Louis region per unit of output of the kth good or economic sector. The $1 \times t$ vector of elements W^k is $\mathbf{W}$.

x_j The activity level of production-abatement process j. The subscript j includes the number of the pollution source followed by a letter, which indicates an abatement process. The lowest cost abatement (or nonabatement) process for a particular source is denoted by "a," the next more costly process by "b," etc.

$\mathbf{x}$ An $n \times 1$ vector of decision variables x_j. The least-cost set of activities is $\mathbf{x}^*$. The vector in which each base year activity x_j is equal to the corresponding source magnitude s_j and all of the alternative activity levels are zero is $\mathbf{x}_s$. The set of activity levels most closely related to legal regulations is $\mathbf{x}^r$, and the set of activities that minimizes job loss subject to an efficiency constraint is $\mathbf{x}^D$.

X_j Activity level of the jth abatement process in the model in which pollution sources are disaggregated according to location in the airshed. This decision variable is an element of an $N \times 1$ column vector $\mathbf{X}$. Because of disaggregation the dimension N exceeds n. The actual activity levels in year t are represented by the vector $\mathbf{X}_t$.

y_k Total physical output of good k or economic sector k per unit of time. The total market value of this output in the absence of air pollution control regulations, given the base-year level of abatement is $(p_k y_k)$. The $t \times 1$ vector whose typical element is y_k is $\mathbf{y}$. The set of outputs in the absence of regulations is the vector $\mathbf{y}^0$. The vector of Pareto optimal outputs of goods is $\mathbf{y}^*$.

y_{ij} Total quantity of good i consumed by the jth household per unit of time. A specific quantity is denoted by $\bar{y}_{ij}$ or $\bar{\bar{y}}_{ij}$.

Δy_k^i The total change in the quantity of good k (or output of sector k) produced or sold as a consequence of the jth control method activity. The $t \times 1$ vector of total output changes as a consequence of the abatement feedback is $\Delta \mathbf{y}$.

Υ The quantity of a composite good. The quantity consumed by the ith household is Υ_i.

Z Total cost of air pollution abatement.

α_i Quantity of intermediate good i required in production per unit of labor input; also quantity of polluting source level i per unit of resource cost. This coefficient is the typical element of the $m \times 1$ vector $\boldsymbol{\alpha}$.

β_k^j The change in the output of good k per unit of the jth control method activity as a consequence of a feedback effect.

ε_k Price elasticity of demand for the kth good or output of the kth economic sector.

θ_j The fraction of the year that the wind direction is through the compass sector in which the jth source is located and toward the receptor station.

θ_{ij} The proportional change in the distortion of the ratio of prices of goods i and j. When the change is distortion reducing, θ_{ij} is negative. When the change is distortion increasing, θ_{ij} is positive.

μ A heuristic measure of the overall impact of a least-cost set of standards on price distortion.

π^i Shadow price of the ith pollutant. The $p \times 1$ vector of pollutant shadow prices is π.

π_i Shadow price of the ith pollution source level.

σ_i The pure production cost of the i intermediate good.

ϕ Pollution or emission fee, such as a Pigouvian fee.

ρ Reduced cost; the unit penalty for using an inefficient process.

Ω Multiplier in the pure-abatement model.

BIBLIOGRAPHY

Abrams, Lawrence W., and James L. Barr, "Corrective Taxes for Pollution Control," *Journal of Environmental Economics and Management*, Volume 1, December 1974, pp. 296–318.

Anderson, Robert J., Jr., and Thomas D. Crocker, "The Economics of Air Pollution," in *Air Pollution and the Social Sciences*, edited by Paul B. Downing. New York: Praeger, 1971, p. 156.

Atkinson, Scott E.; and Donald H. Lewis, "A Cost-Effectiveness Analysis of Alternative Air Quality Control Strategies," *Journal of Environmental Economics and Management*, Volume 1, November 1974, pp. 237–250.

Atkinson, Scott E., and Donald H. Lewis "Determination and Implementation of Optimal Air Quality Standards," *Journal of Environmental Economics and Management*, Volume 3, December 1976, pp. 363–380.

Ayres, Robert U., and Allen V. Kneese, "Production, Consumption, and Externalities," *American Economic Review*, Volume 59, June 1969, pp. 282–297.

Babcock, Lyndon R., Jr., "A Combined Pollution Index for Measurement of Total Air Pollution," *Journal of the Air Pollution Control Association*, Volume 20, October 1970, pp. 653–659.

Baumol, William J., "On the Social Rate of Discount," *American Economic Review*, Volume 5, September 1968, pp. 788–802.

Baumol, William J., "On Taxation and the Control of Externalities," *The American Economic Review*, Volume 62, June 1972, pp. 307–322.

Baumol, William J. and Wallace E. Oates, "The Use of Standards and Prices for Protection of the Environment," in *The Economics of Environment*, edited by Peter Bohm and Allen V. Kneese. London: St. Martin's Press, 1971.

Baumol. William J., *The Theory of Environmental Policy*. Englewood Cliffs: Prentice-Hall, Inc., 1975.

Beneke, Raymond R., and Ronald Winterboer, *Linear Programming Applications to Agriculture*. Ames, Iowa: Iowa University Press, 1973.

Berglas, Eitan, "Pollution Control and Intercommunity Trade," *The Bell Journal of Economics*, Spring 1977, pp. 217–233.

Burton, Ellison S., Edward H. Pechan, and William Sanjour, "Solving the Air Pollution Control Puzzle," *Environmental Science and Technology*, Volume 7, May 1973.

Clarke, J. F., "A Simple Diffusion Model for Calculating Point Concentrations from Multiple Sources," *Journal of the Air Pollution Control Association*, Volume 14, no. 9, 1964, pp.347–352.

Cohen, Alan S., Gideon Fishelson, and John L. Gardner, *Residential Fuel Policy and the Environment*. Cambridge: Ballinger, 1974.

Cooper, Leon, and David Steinberg, *Methods and Applications of Linear Programming*. Philadelphia: W. B. Saunders Company, 1974.

Danielson, John A., *Air Pollution Engineering Manual*. Cincinnati: National Center for Air Pollution Control, 1967.

Deacon, Robert, "The Environment: A Challenge or Public Policy," in *Economic Analysis of Pressing Social Problems*, edited by Llad Phillips and Harold L. Votey, Jr. Chicago: Rand McNally, 1974.

Dewees, Donald N., *Economics and Public Policy: The Automobile Pollution Case*. Cambridge: MIT Press, 1974.

Dick, Daniel T., *Pollution, Congestion, and Nuisance*. Lexington: D. C. Heath & Company, 1974.

Dorfman, Robert, "Incidence of the Benefits and Costs of Environmental Programs," *American Economic Review*, Volume 67, February 1977, pp. 333–340.

Dorfman, Robert, and Nancy S., *Economics of the Environment*. New York: Norton, 1972.

Downing, Paul B., and W. D. Watson, Jr., "The Economics of Enforcing Air Pollution Controls," *Journal of Environmental Economics and Management*. Volume 1, November 1974, pp. 219–236.

Dwyer, Paul S., *Linear Computations*. New York: John Wiley and Sons, 1951.

Emerson, M. Jarvin, and Keith R. Leitner, "Effects of State Pollution Control Programs on Industrial Location," Paper Presented at the Annual Meeting of the Midwest Economics Association, St. Louis, 1976.

Ferrar, Terry A., Peter G. Sassone, and Alan B. Brownstein, "Charges and Political Realities—A Qualification," *Journal of Environmental Systems*, Volume 5, No. 2, 1975, pp. 95–101.

Fisher, Anthony C., and Frederick M. Peterson, "The Environment in Economics: A Survey," *Journal of Economic Literature*, Volume 14, March 1976, pp. 1–33.

Foster, Edward, and Hugo Sonenschein, "Price Distortion and Economic Welfare," *Econometrica*, Volume 38, March 1970, pp. 281–297.

Freeman, A. Myrick, "Distribution of Environmental Quality," in *Environmental Quality Analysis*, edited by Allen V. Kneese and Blair T. Bower. Baltimore: Johns Hopkins, 1972, p. 265.

Goeller, Bruce F., et al., *San Diego Clean Air Project*, Rand Report R-1362-SD, Santa Monica, California, December 1973.

Gipson, Gerald L., Warren Freas, and Edwin L. Meyer, Jr., "Evaluation of Techniques for Obtaining Least-Cost Regional Strategies for Control of Sulfur Dioxide and Suspended Particulates," *Environmental Science and Technology*, Volume 9, April 1975, pp. 354–359.

Green, Marvin H., "An Air Pollution Index Based on Sulfur Dioxide and Smoke Shade," *Journal of the Air Pollution Control Association*, Volume 11, December 1966, pp. 703–706.

Guldmann, Jean-Michel, *Optimization Models for Air Pollution Control Strategies and Location of Industries: A Case Study of the Haifa Region*, Unpublished Master's Degree Thesis, Technion, Haifa, Israel, May 1973.

Hadley, G., *Linear Programming*. Reading: Addison-Wesley, 1962.

Hicks, J. R., *Value and Capital*. London: Oxford University Press, 1965.

Hougen, O. A., K. M. Watson, and R. A. Rogatz, *Chemical Process Principles*, New York: John Wiley, 1962.

Houthakker, H. S., and L. D. Taylor,

Consumer Demand in the United States: Analyses and Projections, 2nd Edition. Cambridge: Harvard University Press, 1970.

Jackson, W. E., and H. C. Wohlers, "Regional Air Pollution Control Costs," *Journal of the Air Pollution Control Association*, Volume 22, September 1972, pp. 679–684.

Kahn, Alfred, Robert B. Rutledge, Gustave L. Davis, Jane A. Altes, George E. Gantner, Charles A. Thornton, and Norval D. Wallace, "Carboxyhemoglobin Sources in the Metropolitan St. Louis Population," *Archives of Environmental Health*, Volume 29, September 1974, pp. 127–135.

Kaldor, Nicholas, "Welfare Propositions of Economics and Interpersonal Comparisons of Utility," *The Economic Journal*, Volume 49, 1939, pp. 549–552.

Kawamata, Kunio, "Price Distortion and Potential Welfare," *Econometrica*, Volume 42, May 1974, pp. 435–460.

Kneese, Allen V., and Blair T. Bower, *Managing Water Quality: Economics, Technology, Institutions*. Baltimore: Johns Hopkins, 1968.

Kneese, Allen V., and Charles L. Schultze, *Pollution, Prices, and Public Policy*. Washington, D. C.: Brookings, 1975.

Kohn, Robert E., "Achieving Air Quality Goals at Minimum Cost," *Washington University Law Quarterly*, Spring 1968, pp. 325–360.

Kohn, Robert E., *A Linear Programming Model for Air Pollution Control*, unpublished doctoral dissertation, St. Louis, Washington University, 1969a.

Kohn, Robert E., "A Mathematical Programming Model for Air Pollution Control," *School Science and Mathematics*, June 1969b, pp. 487–494.

Kohn, Robert E., "Linear Programming Model for Air Pollution Control: A Pilot Study of the St. Louis Airshed," *Journal of the Air Pollution Control Association*, Volume 20, no. 2, February 1970a, pp. 78–82.

Kohn, Robert E., "Statement on Air Quality Standards and the August, 1969, St. Louis Pollution Episode," *Hearings before the Subcommittee on Air and Water Pollution of the Committee on Public Works, United States Senate, Ninety-First Congress, First Session on Problems and Programs Associated with the Control of Air Pollution, October 27, 1969, St. Louis, Missouri.* Washington, D.C.: U. S. Government Printing Office, 1970b, pp. 46–50.

Kohn, Robert E., "Abatement Strategy and Air Quality Standards," in *Development of Air Quality Standards*, edited by A. Atkisson and R. Gaines. Columbus: Charles E. Merrill Publishing Co., 1970c, pp. 103–122.

Kohn, Robert E., "Application of Linear Programming to a Controversy on Air Pollution Control," *Management Science*, Volume 17, no. 10, June 1971a, pp. 609–621.

Kohn, Robert E., "Joint-Outputs of Land and Water Wastes in a Linear Programming Model for Air Pollution Control," *Proceedings of the Social Statistics Section 1970 Annual Meeting*. Washington, D. C.: American Statistical Association, 1971b, pp. 207–214.

Kohn, Robert E., "Air Quality Goals, the Multi-Product Production Function, and the Opportunity Cost of Capital," *Southern Economic Journal*, Volume 38, no. 2, October 1971c, pp. 156–160.

Kohn, Robert E., "Optimal Air Quality Standards," *Econometrica*, Volume 39, no. 6, November 1971d, pp. 983–995.

Kohn, Robert E., "A Cost-Effectiveness Model for Air Pollution Control with a Single Stochastic Variable," *Journal of the American Statistical Association*, Volume 67, March 1972a, pp. 19–22.

Kohn. Robert E., "Price Elasticities of Demand and Air Pollution Control," *Review of Economics and Statistics*, Volume 54, November 1972b, pp. 392–400.

Kohn, Robert E., "Labor Displacement and Air-Pollution Control," *Operations Research*, Volume 21, September-October 1973, pp. 1063–1070.

Kohn, Robert E., "Industrial Location and Air Pollution Abatement," *Journal of the Regional Science Association*, Volume 14, 1974a, pp. 55–63.

Kohn, Robert E., "Comments on Schuler Paper," *Papers of the Regional Science Association*, Volume 32, 1974b, pp. 149–152.

Kohn, Robert E., "Input-Output Analysis and Air Pollution Control," in *Economic Analysis of Environmental Problems*, edited by Edwin Mills, Universities-National Bureau Committee for Economic Research Conference Volume. New York: Columbia University Press, 1975a, pp. 239–272.

Kohn, Robert E., *Air Pollution Control: A Welfare Economic Interpretation*. Lexington: D.C. Health, 1975b.

Kohn, Robert E., "Emission Standards and Price Distortion," *Journal of Environmental Economics and Management*, Volume 4, no. 3, 1977a, pp. 200–208.

Kohn, Robert E., "A Benefit-Cost Analysis to Determine a Population Cutoff for a Statewide Ban on Leaf Burning," *Journal of the Air Pollution Control Association*, Volume 27, September 1977b, pp. 887–889.

Kohn, Robert E., and Donald E. Burlingame, "Air Quality Control Model Combining Data on Morbidity and Pollution Abatement," *Decision Sciences*, July 1971, pp. 300–310.

Kohn, Robert E., and Eric Weger, "Pollution Control by Locational Change and Technological Abatement," *Journal of the Air Pollution Control Association*, Volume 23, December 1973, pp. 1045–1047.

Kohn, Robert E., and Donald C. Aucamp, "Abatement, Avoidance, and Nonconvexity," *American Economic Review*, Volume 66, December 1976, pp. 947–952.

Larsen, Ralph I., "A Method for Determining Source Reduction Required to Meet Air Quality Standards," *Journal of the Air Pollution Control Association*, Volume 11, February 1961, pp. 71–76.

Larsen, Ralph I., "Relating Air Pollutant Effects to Concentration and Control," *Journal of the Air Pollution Control Association*, Volume 20, April 1970, pp. 214–225.

Larsen, R. I., C. E. Zimmer, D. A. Lynn, and K. G. Blemel, "Analyzing Air Pollutant Concentration and Dosage Data," *Journal of the Air Pollution Control Association*, Volume 17, February 1967, pp. 85–93.

Leontief, Wassily, "Environmental Repercussions and the Economic Structure:

An Input-Output Approach," *Review of Economics and Statistics*, Volume 52, August 1970, pp. 262–271.

Leung, Kenneth Ch'uan-k'ai, and Jeffrey A. Klein, *The Environmental Control Industry*. Ann Arbor: Ann Arbor Science, 1976.

Lillis, Edward J., and Jean J. Schueneman, "Continuous Emission Monitoring: Objectives and Requirements," *Journal of the Air Pollution Control Association*, Volume 25, August 1975, pp. 804–809.

Lin, Steven A. Y., *Theory and Measurement of Economic Externalities*. New York: Academic Press, 1976.

Lin, Steven A. Y., Robert E. Kohn, and Donald E. Burlingame, "Air Quality Control Model Combining Data on Morbidity and Pollution Abatement," *Decision Sciences*, Volume 3, April 1972, pp. 144–146.

Liu, Ben-chieh, *Interindustrial Structure of the St. Louis Region, 1967*. St. Louis: St. Louis Regional Industrial Development Corporation, 1968.

Lunche, Robert G. et al., *Air Pollution Engineering in Los Angeles County*. County of Los Angeles: Air Pollution Control District, 1968.

McFarland, D. G., E. V. Barry, and J. W. DeNardo, "The Development of a Quantitative Objective Air Pollution Forecast System for Allegheny County, Pennsylvania," paper presented at the 62nd Annual Meeting, Air Pollution Control Association, New York City, June 22–26, 1969.

Maler, Karl-Goran, *Environmental Economics: A Theoretical Inquiry*. Baltimore: Johns Hopkins Press, 1974.

Meade, James E., *The Theory of Economic Externalities*. Leiden: A. W. Sijthoff, 1973.

Miernyk, William H., and John T. Sears, *Air Pollution Abatement and Regional Economic Development*. Lexington: D. C. Heath and Company, 1974.

Mills, Edwin S., *Urban Economics*. Glenville: Scott, Foresman and Company, 1972.

Nevers, Noel de, "Air Pollution Control Philosophies," *Journal of the Air Pollution Control Association*, Volume 27, March 1977, pp. 197–205.

Oron, Yitzhak, and David Pines, "The Effect of Nuisances Associated with Urban Traffic on Suburbanization and Land Values," *Journal of Urban Economics*, Volume 1, October 1974, pp. 382–394.

Ott, Wayne R., and Gary C. Thom, "A Critical Review of Air Pollution Index Systems in the United States and Canada," *Journal of the Air Pollution Control Association*, Volume 26, May 1976, pp. 460–470.

Page, Talbot, and John Ferejohn, "Externalities as Commodities: Comment," *The American Economic Review*, Volume 64, June 1974, pp. 454–459.

Parsons, D. O., and E. J. Croke, "An Economic Evaluation of Sulfur Dioxide Air Pollution Incident Control," paper presented at the 62nd Annual Meeting, Air Pollution Control Association, New York City, June 22–26, 1969.

Pigou, A. C., *The Economics of Welfare*, 4th edition. London: Macmillan and Company, 1960.

Pooler, F., Jr., "A Prediction Model of Mean Urban Pollution for Use with

Standard Wind Roses," *International Journal of Air and Water Pollution*, Volume 4, 1961, pp. 199–211.

Rader, Trout, "The Welfare Loss from Price Distortions," *Econometrica*, Volume 44, November 1976, pp. 1253–1257.

Ridker, Ronald G., *Economic Costs of Air Pollution*. New York: Praeger, 1967.

Rose, Adam, "The Cost of Stack Gas Cleaning in New York State," *Journal of the Air Pollution Control Association*, Volume 25, October 1975, pp. 1005–1008.

Ruff, Larry E., "The Economic Common Sense of Pollution," *The Public Interest*, Volume 19, Spring 1970, pp. 69–85.

Russell, Clifford, S., and Walter O. Spofford, Jr., "A Regional Environmental Quality Management Model: An Assessment," *Journal of Environmental Economics and Management*, Volume 4, June 1977, pp. 89–110.

Samuelson, Paul A., "The Pure Theory of Public Expenditure," *The Review of Economics and Statistics*, November 1954, pp. 387–389.

Samuelson, Paul A., "Pure Theory of Public Expenditures and Taxation," in *Public Economics*, edited by J. Margolis and H. Guitton. London: MacMillan and Company, 1969, pp. 98–123.

Schneider, Alan M., "An Effluent Fee Schedule for Air Pollutants Based on Pindex," *Journal of the Air Pollution Control Association*, Volume 23, June 1973, pp. 486–489.

Schuler, Richard E., "Air Quality Improvement and Long-Run Urban Form," *Papers of the Regional Science As-sociation*, Volume 32, 1974, pp. 133–148.

Seinfeld, John H., and Chwan P. Kyan, "Determination of Optimal Air Pollution Control Strategies," *Socio-Economic Planning Sciences*, Volume 5, June 1971, pp. 173–190.

Shepard, Donald S., "A Load Shifting Model for Air Pollution Control in the Electric Power Industry," *Journal of the Air Pollution Control Association*, Volume 20, December 1970, pp. 756–761.

Siegel, Richard D., John Ehrenfeld, and Paul Morganstern, "A Strategy for Reduction of Particulate Emissions in the Boston Area," *Journal of the Air Pollution Control Association*, Volume 25, March 1975, pp. 256–259.

Smith, M. E., *International Symposium on Chemical Reactions in the Lower and Upper Atmosphere*, 1961 Advance Papers. San Francisco: Stanford Research Institute, pp. 273–286.

Snee, Ronald D., and John M. Pierrard, "The Annual Average: An Alternative to the Second Highest Value as a Measure of Air Quality," *Journal of the Air Pollution Control Association*, Volume 27, February 1977, pp. 131–133.

Solow, Robert M., "The Economist's Approach to Pollution and Its Control," *Social Science for Winter 1972*, 1972, pp. 15–25.

Sterling, T. D., S. V. Pollack, and J. J. Phair, "Urban Hospital Morbidity and Air Pollution: A Second Report," *Archives of Environmental Health*, Volume 15, September 1967, pp. 362–374.

Teller, Azriel, "Air Pollution Abatement: Economic Rationality and Reality," *Daedalus*, Fall 1967, pp. 1082–1098.

Tietenberg, T. H., "On Taxation and the Control of Externalities: Comment," *The American Economic Review*, Volume 64, June 1974, pp. 462–466.

Tihansky, D. P., "Economic Models of Industrial Pollution Control in Regional Planning, " *Environment and Planning*, Volume 5, 1973, pp. 339–356.

Tillman, Frank A., and E. Stanley Lee, "Episode Control of Air Pollution," *The Journal of Environmental Sciences*, September-October 1975, pp. 21–26.

Trijonis, John C., "Economic Air Pollution Control Model for Los Angeles County in 1975," *Environmental Science and Technology*, Volume 8, September 1974, pp. 811–826.

Turner, D. B., *Workbook of Atmospheric Dispersion Estimates*, National Air Pollution Control Administration, Cincinnati, 1969.

Van Belle, Gerald, and Marvin Schneiderman, "Some Statistical Aspects of Environmental Pollution and Protection," *International Statistical Review*, Volume 41, 1973, pp. 315–331.

Venezia, R., and G. Ozolins, *Interstate Air Pollution Study, Part II, Air Pollution Emissions Inventory*. Cincinnati: National Center for Air Pollution Control, 1966.

Vickery, William, "Theoretical and Partical Possibilities and Limitations of a Market Mechanism Approach to Air Pollution Control," Paper Presented at the Air Pollution Control Association Meetings, Cleveland, June 1967.

Waddell, T., "The Economic Damages of Air Pollution," U.S. Environmental Protection Agency, 1973.

Wanta, R. C., "Meteorology and Air Pollution," in *Air Pollution*, Volume 1, edited by Arthur C. Stern. Academic Press: New York, 1968, pp. 216–217.

Weidenbaum, Murray L., *Government Mandated Price Increases*. Washington: American Enterprise Institute for Public Policy Research, 1975.

Williams, J. D., G. Ozolins, J. W. Sadler, and J. R. Farmer, *Interstate Air Pollution Study, Phase II Project Report: VIII, a Proposal for an Air Resource Management Program*. Cincinnati: National Center for Air Pollution Control, 1967.

Yan, Chiou-Shuang, An-Min Chung, Edward C. Koziara, and Andrew G. Verzilli, "Air Pollution Control Costs and Consumption Pattern Effects," *Journal of Environmental Economics and Management*, Volume 2, September 1975, pp. 60–68.

Zerbe, Richard O., "Theoretical Efficiency in Pollution Control," *Western Economic Journal*, Volume 8, December 1970, pp. 364–376.

Zerbe, Richard O., and Kevin Croke, *Urban Transportation for the Environment*. Cambridge: Ballinger, 1975.

Zimmer, Charles C., and Ralph I. Larsen, "Calculating Air Quality and its Control," *Journal of the Air Pollution Control Association*, Volume 15, December 1965, pp. 570–571.

Air Quality Implementation Planning Program, TRW Systems Group, National Air Pollution Control Administration (PH 22-68-60), Washington, D.C., March 1970.

Cleaning Our Environment: The Chemical Basis for Action, American Chemical Society, Washington, D.C., 1969.

Climatological Data, Missouri Annual Summary, 1970, Environmental Data Service, Volume 74.

Current Industrial Reports, Pollution Abatement: Cost and Expenditures, MA–200(75)–1, Department of Commerce, Bureau of the Census, Washington, D.C., 1977.

Environmental Pollutants and the Urban Economy; Final Report, Chicago: University of Chicago Center for Urban Studies and Argonne National Laboratory, 1976.

Environmental Quality—1976, The Seventh Annual Report of the Council on Environmental Quality, September 1976.

INDEX

www.ingramcontent.com/pod-product-compliance
Lightning Source LLC
Chambersburg PA
CBHW061145220326
41599CB00025B/4365